The Special Theory of Relativity:
Mathematical Refutations

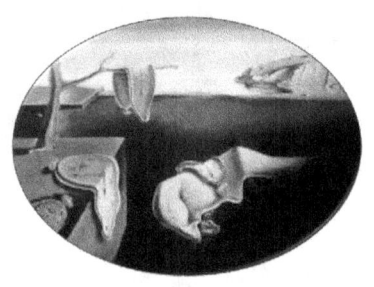

The Special Theory of Relativity: Mathematical Refutations

Radwan M. Kassir

Second Edition

CreateSpace
Charleston

Published in 2014 thru CreateSpace, a DBA of On-Demand
Publishing, LLC
7290 Investment Drive Suite B
North Charleston, SC 29418
www.createspace.com

ISBN-13   978-1500530969
ISBN-10   1500530964

To my father

In questions of science the authority of a thousand is not worth the humble reasoning of a single individual.

— *Galileo Galilei*

To kill an error is as good a service as, and sometimes even better than, the establishing of a new truth or fact.

— *Charles Darwin*

TABLE OF CONTENTS

FOREWORD

Einstein's Special Relativity has been extensively criticized since the time of its first publication in 1905. Doubts on the bases of scientific, mathematical, and philosophical arguments have been expressed. Criticism, on both academic and non-academic levels, has been mainly motivated by the unordinary physical phenomena of the time dilation and length contraction in moving frames, emerging from the purely mathematical formulation of the theory, in addition to resultant numerous paradoxes combined with the inconsistency and ambiguity in their resolutions. Many opponents have shown various inconsistencies in the theory, with valid grounds to topple it. Yet, the theory seems to have been well backed up and protected by the physics community, probably for political and economic considerations!

In order to officially abandon an established scientific theory for being deemed invalid, the concerned scientific authorities must issue in consent a well-documented statement declaring such desertion with tangible justifications. Ironically, refutation of the Special Relativity cannot possibly be emanated from the physics community. The reason is that a physicist's mind is formed to take established physics theories, especially Relativity, for granted. Physicists are systematically educated and brainwashed to the point that this theory becomes an unquestionable, blindly followed belief. They are so unreasonably convinced about the correctness and validity of the "proven" theory, they are appalled at the idea of questioning it, even at considering any challenging ideas or doubtful views; Relativity has been brought for them to the level of a religion!

If a recognized physicist promoted skepticism about Relativity—like the case of Dingle—they will be discredited and expelled from the physics community, implicitly facing the charges of defection and professional incompetence! On the other hand, if a challenge was coming from outside the physics community circle, i.e. from independent thinkers whose profession doesn't belong to the physics establishments (e.g., Beckmann and Kelly), it would be prejudicially considered by the physics community as an unreliable amateur attempt with no real value or impact on the "soundness" of the theory, and therefore ignored, no matter how good or valid that challenge is!

It follows that the theory of relativity will continue to be falsely and unjustly defended and maintained by the biased orthodox authorities of the physics community including recognized universities, scientific institutions and organizations, mainstream journals—whose editors block the publications of any dissident works—and research centers. They ensure the emergence of any work threatening Relativity will be suppressed.

This is reminiscent of the eras in the history of civilizations when wrong scientific beliefs governed and persisted for long times. Eras when the earth was believed to be flat; when the earth was the center of the universe around which the sun and heaven stars revolved; and when scientists and thinkers were to be condemned as sinners had they dared to challenge the prevailing [wrong] beliefs—and many were executed for doing so! Eras when only "divine", dogmatic establishments, ruled by circles of authoritative individuals having the sole intentions of promoting their self-interests, were given the authority to judge evolving scientific ideas and conjectures, and accept only those promoting their own benefits and/or beliefs.

Such scientific domination protecting the relativity theory will remain the status quo, until further convincing, serious researches and studies whose findings undoubtedly disprove the validity of the theory are established, and considered objectively by influential establishments whose directives can affect the academic society as well as the physics community standpoints.

The analytical studies on the Special Relativity presented in this book fall in the category of such serious researches. These studies disprove the theory by the means of concrete mathematical approaches leading to solid evidences of its un-tenability. Promoting such studies would provide a good service to modern physics by urging a quest to put its drifting progress back in the right track again!

PREFACE

Conceivably, Einstein's theory of the Special Relativity (SR) has been the most criticized theoretical physics work of the 20th century. A quick search of the web reveals the substantial number of a diversity of papers, essays, and books launched against it; but why!? The theory itself involves some counterintuitive assumptions leading to absurd, rather unrealistic outcomes and paradoxes, giving the theory a kind of a fictitious, surrealistic aspect!

Most critics argue that the theory is mathematically sound, yet its mathematical formulation is based on faulty assumptions. Therefore, most attempts to disprove the SR have been oriented towards identifying inconsistent or illogical outcomes concluded from its predictions. For instance, the clock paradox has been the subject of a long debate between relativity opponents from one side and supporters from the other side.

In analyzing the SR mathematical formulation, however, it has been revealed to me that the two fundamental SR assumptions are inconsistent with each other, resulting in transformation equations embedding fundamental mathematical contradictions. In addition, the constancy of the speed of light assumption requires some initial conditions that are ignored in the SR transformation equations derivation, which results in hidden mathematical contradictions in these equations. Realizing these inconsistencies, basic violations, and resulting contradictions, initiated the idea of carrying out a detailed analytical study leading to the consistent conclusion of the unviability of SR through many different mathematical approaches.

I have come to the conclusion, through undisputable mathematical evidences, that SR is built upon the interpretations of a theoretical, deceptively plausible coordinate linear transformation—in a hypothetically four-dimensional space-time—derived from not only imaginary assumptions, but contradictory as well!

— Radwan M. Kassir

June 2014

INTRODUCTION

The idea of the Special Relativity theory (SR) as brought up in the famous Einstein's 1905 paper[1] is based on the constancy of the speed of light postulate; that is the speed of light is constant with respect to all inertial frames of reference. This has been categorized as the second postulate of the theory. The first postulate, the principle of relativity, stating that the laws of physics are the same in all inertial frames of reference, was seemingly introduced as an essential tool needed in the mathematical formulation of the SR.

In order to understand the implication of the constancy of the speed of light postulate, let's consider two inertial frames of reference, K and K', in relative translational motion with velocity v, and let (x, y, z) and (x', y', z') be two coordinate systems associated with K and K', in such a way that the corresponding axes are parallel, and v is in the direction of the overlapping x-x' axes. Let c be the speed of light relative to K. If, at the instant of time when the two frames are coinciding, a light ray is emitted from a point at the origins in the x-x' direction, then according to the classical Galilean transformation, after an interval of time t has elapsed, the light ray will have travelled the distances ct and $(c - v)t$ with respect to K and K', respectively.

On the other hand, the constancy of the speed of light assumption requires that the speed of light be the same with respect to both frames, which would make the distance travelled by the light ray equal to ct in both frames, imposing the impossibility $ct = ct \pm vt$ for $v \neq 0$, necessitating a time transformation, for instance from t to t' in K' with respect to K. Thus, the distances travelled by the light ray become ct and ct' in K and K', respectively, with ct' being shorter than ct from the perspective of K. Yet, the time and distance travelled in K' remain unchanged from the perspective of K'. Therefore, the fact that the distance travelled by the light ray in K' is perceived to be shorter with respect to K, a spatial transformation has also occurred in K' relative to K. By the symmetry assumption—which will be shown to be inadequate—implicated by the SR second postulate, and according to the SR first postulate, similar time and space transformations should occur in K with respect to K'. Hence, we see that the space and time must be deformed in order to satisfy the speed of light constancy assumption.

It follows that, as a consequence of the speed of light constancy principle, the time and space dimensions become relative entities depending on the relative motion of the inertial reference frame in which they're measured, and the classical Galilean transformation relating between the coordinates of two inertial frames in relative motion becomes inapplicable for reference frames moving at high relativistic speeds.

A general relativistic transformation needs then to be established. The derivation of such transformation, its interpretation, and its predictions constituted the main theme of the SR. The Transformation was named after the Dutch physicist Hendrik Lorentz, for his earlier works on the same transformation in connection with justifying the null results of the famous Michelson-Morley experiment,[2] in an attempt to save the ether theory. Einstein's interpretations of the transformation, however, threw the ether model, and established new concepts of space and time. Physical lengths in the "traveling" reference frame are contracted in the direction of relative motion with respect to the "stationary" reference frame, whereas the time in K' is dilated with respect to the "stationary" frame (i.e., time runs slower in the traveling frame relative to the stationary frame).

The predicted length contraction and time dilation values were first physically deduced from the constancy of the speed of light principle. This can be done through calculating the travel times of a light ray traveling the same round trip distance $2L$ in the longitudinal and transverse directions in the "traveling" reference frame K' with respect to K. It was shown that the longitudinal travel time ($\gamma^2 2L / c$) was scaled by a factor of $\gamma = \left(\sqrt{1 - v^2 / c^2}\right)^{-1}$ relative to the transverse travel time ($\gamma 2L / c$). Since, in line with the SR second postulate, the travel time must be the same in the longitudinal and transverse directions, the travel length in the longitudinal direction—being the relative motion direction— must be contracted by the same factor, γ, so that the longitudinal travel time would be contracted from $\gamma^2 2L / c$ to $\gamma 2L / c$. When compared to the calculated travel time ($2L / c$) in K', the travel time in K is found to be dilated by the factor γ. Therefore, time dilation was regarded as an effect of the relative motion, even though the actual consequence of the length contraction (assumed effect of the relative motion) should be indeed a time contraction as well (from $\gamma^2 2L / c$ to $\gamma 2L / c$)!

It follows that a light clock in K' "ticking" at a period of $2L / c$ with respect to K', would be "ticking" at a dilated period of $\gamma 2L / c$ with respect to K. Whereas the length of an object in K' would be contracted with respect to K by the same factor, γ.

This physically deduced transformation of space and time needs to be mathematically reconciled, which couldn't be possible, since the resulting length contraction with respect to K would inevitably lead to a time contraction as well. In fact, a close examination of the mathematically derived transformation (Lorentz transformation) based on the speed of light postulate, actually reveals a time contraction relative to K, misinterpreted as a time dilation in the course of the SR.

The speed of light postulate is therefore, on top of its absurdity, inconsistent, and leads to contradictory transformation equations, the basis of the SR predictions. Mathematical contradictions resulting from these equations are demonstrated through different approaches presented in details in this book's six chapters.

Chapter 1 considers the Michelson-Morley experiment, and the introduction of the Fitzgerald contraction as a physical justification of the experiment null result. The physical deduction of a time dilation is demonstrated to be mathematically unreconciled.

Chapter 2 consists of thorough mathematical analyses of the Lorentz transformation (LT) equations and the speed of light postulate. In the SR derivation of the LT, the initial conditions required by the constancy of the speed of light are ignored in the respective developed equation. The resulting LT equations are demonstrated to lead to mathematical contradictions, and the speed of light principle is deemed to be unviable.

Chapter 3 carries out a simplified derivation of the LT from the constancy of the speed of light principle and the Galilean transformation, identifying some coordinate restrictions. The LT equations are shown to generate mathematical contradictions when they are applied to restricted coordinates. These restricted coordinates, being essentially required to conclude the

SR predictions of time dilation and length contraction, make such predictions unattainable.

Chapter 4 reveals an inconsistency between the constancy of the speed of light and the principle of symmetry emerging from the principle of relativity. It is demonstrated that the speed of light postulate is incompatible with the symmetry principle. When symmetry is imposed on the speed of light principle, the Lorentz transformation equations are obtained, yet with an emerging contradiction requiring the relative motion velocity to be zero.

Chapter 5 is a general study case demonstration of the unviability of the SR prediction of length contraction. It is shown through thought experiments that such prediction is inconsistent with the principle of relativity itself, the SR first postulate. Many physics laws, isotropic with respect to the "traveling" frame, are shown to be, according to SR, direction-dependent with respect to the "stationary" frame, thus violating the SR principle of relativity stating that the physics laws must be the same in all inertial reference frames!

Chapter 6, introduced in the second edition, comprises a straightforward event analysis leading to the reconfirmation of the findings of chapters 2 and 3. The identified essential error in the special relativity formulation—and the suspected hoax used to conceal it—is reinforced with solid evidence, based on the physical interpretation of some temporal event perceptions in the reference frames.

Chapter 1

ON THE INFERENCE OF MICHELSON-MORLEY EXPERIMENT: FITZGERALD CONTRACTION IMPLICATION ON THE TIME DIMENSION

The implication of the Fitzgerald length contraction hypothesis on the time dimension is considered. Originally set as an ad hoc interpretation of the Michelson-Morley experiment null result, the hypothesis is expressed in terms of a space transformation equation inferred from the Galilean transformation, leading to a time conversion exhibiting a contractive property, contradicting the special relativity predictions.

BACKGROUND

The Michelson-Morley experiment[2] was designed in the late 19th century to detect the ether (a conjectured light propagation medium) 'wind' created by the earth motion through the ether-filled space. As light was supposed to travel at a constant speed with respect to the ether, the relative speed of light with respect to earth would then depend on the light propagation direction with respect to the ether 'wind' direction. Fig. 1.1 illustrates the experiment principle. A light beam is sent to a semi silvered mirror placed at 45° angle to the beam direction, splitting it into two beams with directions perpendicular to each other. Each of the two split beams will then travel a distance L from the splitter before it is reflected back to it, and recombining with the other reflected beam in an eyepiece, producing an interference pattern. If the earth is moving through the ether, it would create an ether 'wind' blowing in the opposite direction to its motion, thus delaying the back-and-forth trip of the beam traveling longitudinally to the ether 'wind', with respect to the beam with the transverse motion. This time delay will cause the recombined beams to be out phase, thus a shift in the fringes from the position that would be expected under symmetrical beam trips was anticipated. However, no such shift was observed, even with much more sophisticated variations of the experimental setting providing very high accuracy of the measurements.

If the speed of light with respect to the ether is given as C, and the earth relative velocity as V, then it can be shown that the total longitudinal travel time can be expressed as (derivation will be subsequently presented):

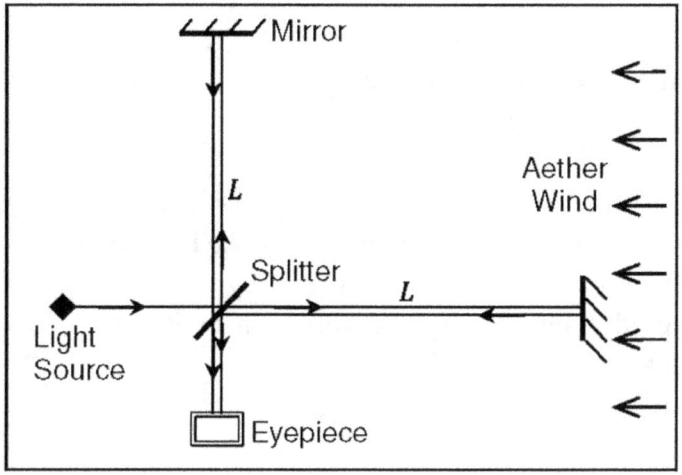

Fig. 1.1 Michelson-Morley experiment setting

$$T_l = \frac{2L}{C} \frac{1}{1 - \dfrac{V^2}{C^2}} \, .$$

Whereas, the total travel time for the transverse beam was $T_t = 2L/C$, as originally indicated by Michelson. Thus T_l is greater than T_t by a factor of $1/(1 - V^2/C^2)$. However, this factor was reduced as a corrected transverse travel time of

$$T_t = \frac{2L}{C} \frac{1}{\sqrt{1 - \dfrac{V^2}{C^2}}}$$

was introduced by Lorentz, taking into consideration the light beam drifting velocity. However, this new time difference still couldn't be reconciled, as the experiment exhibited null result in terms of fringe shift.

In an attempt to resolve this discrepancy, a length contraction hypothesis was proposed by FitzGerald[3] and Lorentz.[4] According to this hypothesis, an object would contract along the direction of its motion by a factor of $1 / \gamma$, with

$$\gamma = \frac{1}{\sqrt{1 - \dfrac{V^2}{C^2}}}$$

being the Lorentz factor. It follows that, the light beam will end up traveling back and forth the contracted longitudinal distance L / γ, and the longitudinal travel time becomes:

$$T_l = \frac{2L / \gamma}{C} \frac{1}{1 - \dfrac{V^2}{C^2}} = \frac{2L}{C} \frac{1}{\sqrt{1 - \dfrac{V^2}{C^2}}}$$

which is the same as the transverse travel time, thus justifying the Michelson-Morley null result.

In this chapter, the implication of the Fitzgerald contraction hypothesis on the time dimension is considered.

LORENTZ FACTOR – PHYSICAL PERSPECTIVE

Starting back from the ether theory and the Michelson-Morley experiment null result, the Lorentz factor is considered in the context of a physical overview, prior to attempting a mathematical reconciliation formulation.

In a certain setting (Fig. 1.2a), where the ether is

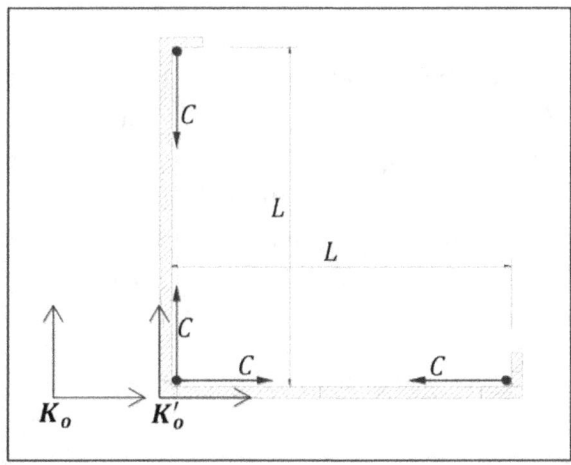

Fig. 1.2a Velocity diagram w.r.t. K_o : Ether is assumed to be totally dragged—No relative motion between ether and earth.

assumed to be totally dragged by the earth, a light beam, having a velocity C with respect to the ether, is to travel a total round trip distance of $2L$ (L being the length of each of the two orthogonal arms of the utilized apparatus), with respect to the earth. Let K_o be a stationary frame of reference with respect to the ether, and K'_o be the earth reference frame; there is no relative motion between K_o and K'_o. The time it takes the light beam to complete the round trip, in either longitudinal or transversal arm direction, as measured by an observer in K_o, or K'_o, will be

$$T_o = \frac{2L}{C}. \qquad (1\text{-}1)$$

In a different setting (Fig. 1.2b), the earth is moving through the ether at a constant speed V. An ether 'wind' of speed V will be thus created with respect to the earth. Two light beams are considered. One beam is to travel a total round trip ground distance of $2L$, going back and forth along the direction of the earth motion.

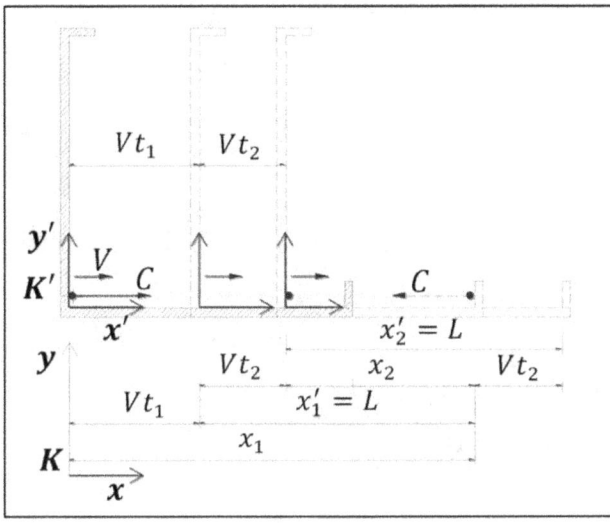

Fig. 1.2b Longitudinal velocity diagram w.r.t. K: Earth is traveling at velocity V w.r.t. the ether— K' is in relative motion w.r.t. K.

A similar round trip in the transverse direction is to be travelled by the other beam. Let $K(x,y,z)$ be a frame of reference at rest with respect to the ether, and $K'(x',y',z')$ be a reference frame attached to earth; K and K' are in relative motion with velocity V.

Longitudinal Travel Time

According to the ether theory, the velocity of the light beam with respect to K (the ether frame) is equal to C. Referring to Fig. 1.2b, we can write

$$x_1 = Vt_1 + x_1', \qquad \text{(1-2)}$$

and

$$x_2 = -Vt_2 + x_2', \qquad \text{(1-3)}$$

where t_1 and t_2 are the forward and backward longitudinal travel time, respectively. Substituting $x_1 = Ct_1$, and $x_2 = Ct_2$, in Eqs. (1-2) and (1-3), and solving for t_1 and t_2, the total round trip time T_l will be determined as

$$T_l = \frac{L}{C - V} + \frac{L}{C + V};$$

$$T_l = \frac{2L/C}{1 - \dfrac{V^2}{C^2}}. \qquad \text{(1-4)}$$

For a stationary observer in K', the ether is 'running' at velocity V in the longitudinal direction, and the light beam upstream and downstream velocities are $C - V$ and $C + V$, respectively, according to the Galilean velocity transformation. Hence, the longitudinal travel time with respect to an observer in K' will be also given by Eq. (1-4).

Transverse Travel Time

For a stationary observer in K', the ether is relatively 'flowing' at speed V in the longitudinal direction. The transverse light beam is traveling in the y' direction with respect to K', at velocity $\vec{C}$ with respect to K. Using the Galilean velocity transformation, the light beam relative velocity $\overrightarrow{C'}$ with respect to K' can be expressed by the following vector addition (Fig. 1.2c).

$$\overrightarrow{C'} = \vec{C} - \vec{V}. \qquad (1\text{-}5)$$

Therefore,

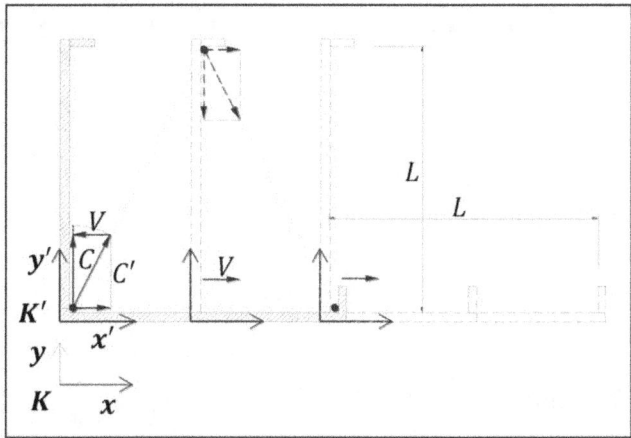

Fig. 1.2c Transverse velocity diagram w.r.t. K: Earth is traveling at velocity V w.r.t. the ether— K' is in relative motion w.r.t. K.

$$C' = \sqrt{C^2 - V^2},$$

or

$$C' = C\sqrt{1 - \frac{V^2}{C^2}}. \qquad (1\text{-}6)$$

It follows that the transverse round trip travel time can be expressed as,

$$T_t = \frac{2L}{C'} = \frac{2L/C}{\sqrt{1 - \dfrac{V^2}{C^2}}}. \qquad (1\text{-}7)$$

Alternatively, with respect to K, the light beam one way transverse distance Ct_1 can be expressed as,

$$C^2 t_1^2 = L^2 + V^2 t_1^2,$$

yielding

$$t_1 = L / \sqrt{C^2 - V^2}.$$

Therefore,

$$T_t = \frac{2L/C}{\sqrt{1 - \dfrac{V^2}{C^2}}},$$

returning Eq. (1-7).

In either approach, the resulting travel time is expressed as the ratio of the arm length to the relative velocity of the light beam with respect to K'.

Length Contraction Hypothesis

In order to validate the ether 'wind' conjecture, following Michelson-Morley null result, the longitudinal and the transverse travel time, T_l and T_t, must be equal. This could be made possible if a space-time modifying transformation was assumed. In fact, comparing Eqs. (1-4) and (1-7), the longitudinal travel time is scaled with respect to the transverse time by a factor of

$$\gamma = \frac{1}{\sqrt{1 - \dfrac{V^2}{C^2}}}. \qquad (1\text{-}8)$$

It is then postulated that the length of a moving object would be contracted along its motion direction by a factor of $1/\gamma$. When this principle is applied in our case to the moving arm of length L, traveling with respect to K at speed V, the longitudinal travel time of the light beam would become just equal to the transverse time. In fact, Eq. (1-4) becomes,

$$T_l = \frac{\dfrac{2L}{C}\dfrac{1}{\gamma}}{\left(1 - \dfrac{V^2}{C^2}\right)} = \frac{\dfrac{2L}{C}\dfrac{1}{\gamma}}{\dfrac{1}{\gamma^2}} = \frac{2L}{C}\gamma = \frac{2L/C}{\sqrt{1 - \dfrac{V^2}{C^2}}}. \qquad (1\text{-}9)$$

It follows from Eqs. (1-1), (1-4), and (1-9) that,

$$T_l = T_t = \gamma T_o, \qquad (1\text{-}10)$$

and

$$L_c = \frac{L}{\gamma}, \qquad (1\text{-}11)$$

where L_c is the contracted length with respect to K, and γ is the Lorentz factor.

Eqs. (1-10) and (1-11) can be interpreted as the time dimension is dilated, and the length's is contracted with respect to the ether frame, due to the earth relative motion, with γ being the time dilation, and $1/\gamma$ the length contraction factors. Yet, in actuality, the Fitzgerald contraction causes the longitudinal travel time to contract from $\gamma^2(2L/C)$ to $\gamma(2L/C)$ —but still dilated by a factor of γ with respect to

$T_o = 2L / C$. Whether this is a valid interpretation of an actual time dilation will be evaluated later in the Mathematical Perspective section.

SPECIAL RELATIVITY'S INTERPRETATION

In the special relativity, the ether conjecture was abandoned, and replaced by the principle of the constancy of the speed of light in all inertial frames of reference, as postulated by Einstein.[1] In contrast with the ether theory, this principle is in fact comparable to viewing the "ether" as if it were at rest with respect to any inertial frame of reference, which makes the speed of light constant (equal to C) with respect to any corresponding observer. It follows that, the particular studied case of the relative motion of the ether-earth reference frames can be generalized to any pair of reference frames in relative motion with any relative velocity less than C, with the only difference being the rest state of the "ether" with respect to an observer in K', which brings the speed of light in K' to C. Hence, the reference frame K', which is in relative motion with respect to the 'stationary' frame K, becomes equivalent to K_o, where the light round trip travel time is $2L / C$ for both the longitudinal and transverse directions.

Therefore, as a consequence of the special relativity postulate, Eqs. (1-10) and (1-11) reduce to

$$T = \gamma T' = \frac{2L'/C}{\sqrt{1 - \dfrac{V^2}{C^2}}}, \qquad (1\text{-}12)$$

and

$$L = \frac{L'}{\gamma} = L'\sqrt{1 - \frac{V^2}{C^2}}. \qquad (1\text{-}13)$$

Where T and T' are the travel time, L and L' the arm length, with respect to K and K', respectively.

Now, Eq. (1-12) can be written as,

$$\Delta t = \gamma \Delta t',$$

or

$$t - t_o = \gamma\left(t' - t'_o\right), \qquad (1\text{-}14)$$

where t_o is a reference time point on the t-axis in K, and t'_o is the corresponding time coordinate in K'. If t_o and t'_o were chosen to be the time coordinates of the origins of K and K', respectively, they can be set to zero, had we assumed that at $t_o = 0$ and $t'_o = 0$, K and K' are coinciding. It follows from Eq. (1-14) that, from the perspective of the frame origins, the K time coordinate with respect to that of K' can be stated as,

$$t = \gamma t', \qquad (1\text{-}15)$$

interpreted as a time dilation with respect to K.

MATHEMATICAL PERSPECTIVE

Since the obtained travel time in the reference frames K and K' seems to involve time transformation, the time coordinate should be introduced to the reference frames. Thus, K and K' are now represented as $K(x, y, z, t)$ and $K'(x', y', z', t')$.

With respect to an observer in K, the hypothesized Fitzgerald length contraction can be expressed by the equation

$$x = Vt + \frac{x'}{\gamma}, \qquad (1\text{-}16)$$

inferred from the Galilean transformation. Eq. (1-16) can be rearranged to the following transformation expression.

$$x' = \gamma\left(x - Vt\right). \qquad (1\text{-}17)$$

Applying the space transformation given by Eq. (1-17), the x-coordinate of the origin of $K(x = 0)$ has a transformed x'-coordinate of

$$X' = -\gamma Vt \qquad (1\text{-}18)$$

at time t (Fig. 1.3a). Similarly, applying the same transformation Eq. (1-17), the x-coordinate of the origin of K',

$$X = Vt, \qquad (1\text{-}19)$$

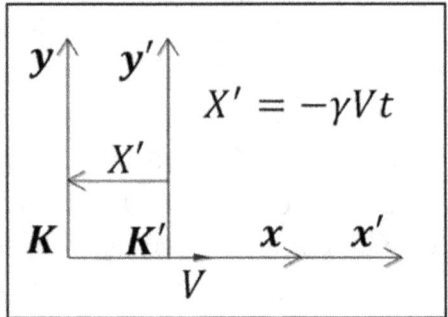

Fig.1.3a x'-coordinate (X') of the origin of K .

has a transformed x'-coordinate of $x' = 0$ (the origin of K') at the same instant of time t (Fig. 1.3b), with respect to K; i.e. for $\Delta t = 0$ ($x = Vt$ and $x' = -\gamma Vt$ are simultaneous events with respect to K), the distance between the frame origins is expressed by Eqs. (1-18) and (1-19) from the perspective of K.

It follows from Eqs. (1-18) and (1-19) that

$$X = -\frac{X'}{\gamma},$$

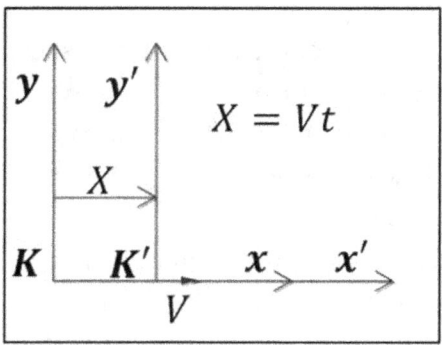

Fig.1.3b x-coordinate (X) of the origin of K'.

which indicates a 'distance' contraction with respect to K (i.e., with respect to an observer in K, the travelled distance X by the K' origin at a certain time instant t, is contracted relative to the absolute value of the corresponding K origin coordinate X' relative to K', attained at the same instant of time t). However, this distance scaling is not in line with the Fitzgerald hypothesis physical interpretation (i.e., the contraction, in the direction of the relative motion, of a moving length interval).

On the other hand, since K is traveling at velocity $-V$ relative to K', then the distance X' (corresponding to $x = 0$) must be equal to $-Vt'$.

Substituting X' in Eq.(1-18), valid for $x = 0$, we get $-Vt' = -\gamma Vt$, yielding

$$t = \frac{t'}{\gamma},$$

which is a time contraction with respect to K for $x = 0$ (i.e., with respect to an observer at the K origin, the time it takes K' origin to travel a certain distance X with respect to K, corresponding to the absolute value of the attained K origin coordinate X' with respect to K', is contracted relative to the corresponding time t' in K').

It follows that he Fitzgerald contraction, expressed by Eq. (1-17), mathematically results in a time contraction with respect to K, which is not in line with the physically derived Eqs. (1-10) and (1-15), interpreted as exhibiting a time dilation with respect to K, for the hypothesized length contraction.

CONCLUSION

For two reference frames relatively moving at a uniform velocity, it is shown that the Fitzgerald contraction hypothesis can be physically interpreted to imply dilation of the time dimension with respect to the stationary frame. Whereas, the hypothesis mathematical formulation results in a space contraction transformation exhibiting time contraction. Hence, the physically anticipated time dilation of the Fitzgerald contraction is not mathematically reconciled.

Chapter 2

LORENTZ TRANSFORMATION: CRITICAL MATHEMATICAL ANALYSES AND FINDINGS

In this chapter, the Lorentz transformation equations are closely examined in connection with the constancy of the speed of light postulate of the special relativity. This study demonstrates that the speed of light postulate is implicitly manifested in the transformation under the form of space-to-time ratio invariance, which has the implication of rendering the frames of reference origin-coordinates undetermined with respect to each other. Yet, Lorentz transformation is shown to be readily constructible on the basis of this conflicting finding. Consequently, the formulated Lorentz transformation is deemed to generate mathematical contradictions, thus defying its tenability. A rationalization of the isolated contradictions is then established, reconfirming the revealed conflicts. An actual interpretation of the Lorentz transformation is presented, demonstrating the unreality of the space-time conversion property attributed to the transformation.

BACKGROUND

The well-known Lorentz transformation, named after the Dutch physicist Hendrik Lorentz, is a set of equations relating the space and time coordinates of two inertial reference frames in relative uniform motion with respect to each other, so that coordinates can be transformed from one reference frame to another. Length contraction and time dilation are supposedly the principal outcome of the Lorentz transformation. Originally, Lorentz developed the transformation to explain, with other physicists (Larmor, Fitzgerald, and Pointcaré), how the speed of light seemed to be independent of the reference frame, following the puzzling results of the famous Michelson-Morley experiment.[2] These equations formed later the basis of Einstein's special relativity. Einstein[1,5] derived Lorentz transformation on the basis of two postulates: 1 – the principle of relativity (i.e., the equations describing the laws of physics have the same form in all proper frames of reference), and 2 – the principle of the constancy of the speed of light in all reference frames.

Einstein's theory of special relativity has received much criticism.[6-11] Doubts on the bases of scientific, mathematical, and philosophical contentions have been expressed. Criticism, on both academic and non-academic levels, has been mainly motivated by the unordinary physical phenomena of the time dilation and length contraction of moving objects, emerging from the purely mathematical formulation of the theory, in addition to numerous paradoxes combined with the inconsistency and ambiguity in their resolutions.[12]

In this chapter, the Lorentz transformation, along with the special relativity speed of light constancy principle used in its derivation, is thoroughly examined in an attempt to reach rational conclusions regarding its ever questioned tenability. Pure mathematical analysis and geometrical tools are used as the main arguments in achieving the objective of this study.

LORENTZ TRANSFORMATION

Consider two inertial frames of reference, $K(x,y,z,t)$ and $K'(x',y',z',t')$, in translational relative motion with parallel corresponding axes, and let their origins be aligned along the overlapped x- and x'-axes. Let v be the relative motion velocity. K and K' are assumed to be overlapping at the time $t = t' = 0$. The space and time coordinates of K and K' are then interrelated by the Lorentz transformation equations given in their present form by Poincaré[13] and subsequently by Einstein[1,5] as follows:

$$x' = \gamma\left(x - vt\right)$$
$$t' = \gamma\left(t - \frac{vx}{c^2}\right) \qquad (2\text{-}1)$$

$$x = \gamma\left(x' + vt'\right)$$
$$t = \gamma\left(t' + \frac{vx'}{c^2}\right) \qquad (2\text{-}2)$$

$$y' = y$$
$$z' = z \qquad (2\text{-}3)$$

$$\gamma = \frac{1}{\sqrt{1 - \dfrac{v^2}{c^2}}} \qquad (2\text{-}4)$$

The above equations are equally applicable for event spatial and temporal intervals. In other words, the coordinates can be expressed as $\Delta x,\ldots,\Delta t$ and $\Delta x',\ldots,\Delta t'$ in the equations.

Eqs. (2-1) and (2-2) result in the following relativistic velocity transformation equations:

$$u' = \frac{u - v}{1 - \dfrac{uv}{c^2}} \left. \vphantom{\begin{array}{c} \\ \\ \\ \\ \\ \end{array}} \right\}$$

$$u = \frac{u' + v}{1 + \dfrac{u'v}{c^2}}$$

(2-5)

where c is the speed of light propagation in empty space, and u and u' are the velocity of a moving body in the x-direction, when measured with respect to K and K', respectively.

It is to be noted that Eq. (2-4) requires that v be smaller than c. Also, Eqs. (2-5) limit the values of u and u' to c (i.e., if $u = c$, then u' is brought to c as well, and vice versa).

LORENTZ TRANSFORMATION ANALYSIS

Constancy of the Speed of Light

Consider two inertial reference frames $K\left(x,y,z,t\right)$ and $K'\left(x',y',z',t'\right)$ moving relative to each other with a uniform velocity v, and suppose at an instant of time $t_o = t'_o = 0$, the frames are overlying. Let a light ray be emitted at this time from the point of the coinciding origins in an arbitrary direction. At time t in K, corresponding to time t' in K' the position vector associated with the light ray will acquire the space-time coordinates (x,y,z,t) and (x',y',z',t') in K and K', respectively. In line with the special relativity constancy of the speed of light principle,[5] the light ray position vector coordinates shall satisfy the following equations, referred to as the light sphere equation transformation.

$$x^2 + y^2 + z^2 = c^2 t^2 \qquad (2\text{-}6)$$

$$x'^2 + y'^2 + z'^2 = c^2 t'^2 \qquad (2\text{-}7)$$

Subtracting Eq. (2-7) from Eq. (2-6), given that the y and z coordinates remain unaltered, leads to the basic constancy of the speed of light equation

$$x^2 - x'^2 = c^2 t^2 - c^2 t'^2. \qquad (2\text{-}8)$$

Considering the Lorentz transformation Eqs. (2-1), it is obvious that, with respect to K, the time t' it takes a light signal, emitted from the point of the coinciding origins at $t = t' = 0$, to travel a distance x' in K' is equal to the time t for the signal to travel the corresponding distance x in K less the signal travel time of the distance vt travelled by the origin of K' at the time t, corrected by the relativistic factor γ. In other words, an event occurring in K' [origin] at the time t with respect to K has already occurred at the time t' equal to t less the signal time of travel from the position of K' [origin] at the time t to K origin, corrected by the relativistic factor γ. Therefore, the term vx / c^2 in the Lorentz time transformation Eq. (2-1) must be the [uncorrected] time it takes the light signal to travel the distance between the origins at the time t with respect to K, or

$$\frac{vx}{c^2} = \frac{vt}{c},$$

leading to

$$x = ct. \qquad (2\text{-}9)$$

Similarly, the Lorentz transformation Eq. (2-2) leads to

$$x' = ct'. \qquad (2\text{-}10)$$

The expressions (2-9) and (2-10) are then an intrinsic part of the Lorentz transformation equations, which makes the expressions

$$x^2 = c^2 t^2 \qquad (2\text{-}11)$$
$$x'^2 = c^2 t'^2 \qquad (2\text{-}12)$$

the inherent solution of the constancy of the speed of light Eq. (2-8), which will equally be revealed in the following demonstration.

Indeed, the Lorentz transformation Eqs. (2-1) can lead to

$$x'^2 = \gamma^2 \left(x^2 + v^2 t^2 - 2xvt \right), \qquad (2\text{-}13)$$

and

$$c^2 t'^2 = \gamma^2 \left(c^2 t^2 + \frac{v^2 x^2}{c^2} - 2xvt \right). \qquad (2\text{-}14)$$

Eliminating the term $2xvt$ from Eqs. (2-13) and (2-14), yields

$$x^2 + v^2 t^2 - \frac{x'^2}{\gamma^2} = c^2 t^2 + \frac{v^2 x^2}{c^2} - \frac{c^2 t'^2}{\gamma^2} . \qquad (2\text{-}15)$$

Similarly, Lorentz transformation Eqs. (2-2) bring about the following expression;

$$-x'^2 - v^2 t'^2 + \frac{x^2}{\gamma^2} = -c^2 t'^2 - \frac{v^2 x'^2}{c^2} + \frac{c^2 t^2}{\gamma^2} . \qquad (2\text{-}16)$$

Adding Eqs. (2-15) and (2-16) will lead to the following expression;

$$x^2\left(1+\frac{1}{\gamma^2}\right) - x'^2\left(1+\frac{1}{\gamma^2}\right) + v^2\left(t^2 - t'^2\right) =$$

$$= c^2t^2\left(1+\frac{1}{\gamma^2}\right) - c^2t'^2\left(1+\frac{1}{\gamma^2}\right) + \frac{v^2}{c^2}\left(x^2 - x'^2\right);$$

which can be simplified to

$$\left(x^2 - x'^2\right)\left(1+\frac{1}{\gamma^2} - \frac{v^2}{c^2}\right) = c^2\left(t^2 - t'^2\right)\left(1+\frac{1}{\gamma^2} - \frac{v^2}{c^2}\right);$$

$$x^2 - x'^2 = c^2\left(t^2 - t'^2\right), \qquad (2\text{-}17)$$

returning actually the speed of light principle Eq.(2-8); thus validating Eqs. (2-15) and (2-16) from the perspective of the special relativity. It should be noted that Eq. (2-17)—or (2-8)—is thus obtained from the Lorentz transformation equations without any restriction on the value of γ (i.e., γ can be replaced in the Lorentz transformation equations by an arbitrary expression, while Eq. (2-17) can still be obtained from the invalid resulting equations).

Whereas, the subtraction of Eq. (2-16) from Eq. (2-15) results in

$$\left(x^2 + x'^2\right)\left(1-\frac{1}{\gamma^2} - \frac{v^2}{c^2}\right) = c^2\left(t^2 + t'^2\right)\left(1-\frac{1}{\gamma^2} - \frac{v^2}{c^2}\right).$$

Now, if we assume for the time being the following equality (as suggested by the above equation),

$$x^2 + x'^2 = c^2 \left(t^2 + t'^2 \right), \qquad (2\text{-}18)$$

then Eqs. (2-17) and (2-18) will readily reduce to Eqs. (2-11) and (2-12), namely

$$x^2 = c^2 t^2$$
$$x'^2 = c^2 t'^2$$

which satisfy both Eqs. (2-15) and (2-16), as well as Eq. (2-8)—when x^2 and x'^2 are replaced with $c^2 t^2$ and $c^2 t'^2$, respectively—thus validating Eq. (2-18) that can also be derived from its consequent Eqs. (2-11) and (2-12).

It should be noted that Eqs. (2-11) and (2-12) can be evidently inferred from Eqs. (2-15) and (2-16).

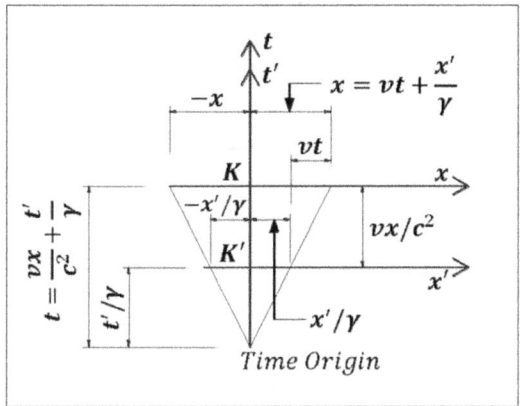

Fig. 2.1a Graphical representation of Lorentz
transformation Eqs. (2-1).

Another practical verification of Eq. (2-11) can be
implemented through Fig. 2.1a depicting a graphical
representation of Lorentz transformation Eqs. (2-1)
(i.e., from the perspective of K) written in the form
shown in the figure, where K and K' are shown
traveling along the overlapped time axes, t and t',
capturing the relative position of the K' coordinates
with respect to those of K.—We note that for the
particular case of $x = ct$, the segment with a slope c
joining the time origin to a point of coordinate x on
the x-axis intersects the x'-axis at a point of
coordinate $x' / \gamma = ct' / \gamma$ (same slope c) with
respect to K, hence the alignment of the time origin,
the point of coordinate x on the x-axis, and the
corresponding point of coordinate x' (x' / γ with

respect to K) on the x'-axis. Using the similar triangles within the graph, we can write,

$$\frac{vt}{vx \ / \ c^2} = \frac{x}{t},$$ (2-19)

yielding, $t^2 = x^2 \ / \ c^2$, or $x^2 = c^2 t^2$.

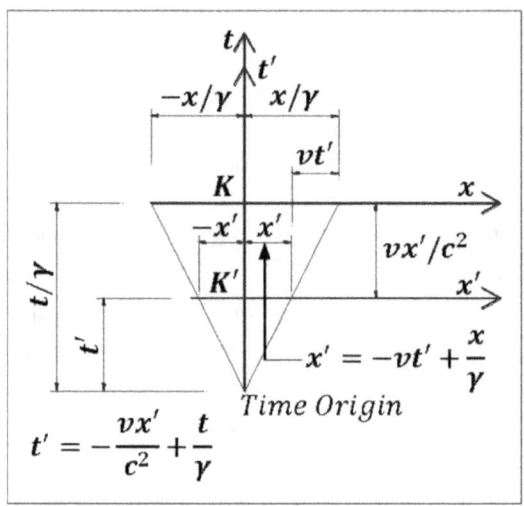

Fig. 2.1b Graphical representation of Lorentz transformation Eqs. (2-2).

Similarly, Eq. (2-12) can be verified using Fig. 2.1b showing a graphical representation of Lorentz transformation Eqs. (2-2) (i.e., from the perspective of K'), as follows;

$$\frac{vt'}{vx'/c^2} = \frac{x'}{t'}, \qquad (2\text{-}20)$$

yielding, $t'^2 = x'^2 / c^2$, or $x'^2 = c^2 t'^2$.

Therefore, the Lorentz transformation equations $x'(x,t)$, $t'(x,t)$, $x(x',t')$, and $t(x',t')$, applicable to any point (x',y',z',t') in K' corresponding to (x,y,z,t) in K, lead simultaneously to the expressions $x^2 = c^2 t^2$ and $x'^2 = c^2 t'^2$, which therefore constitute the distinctive solution of the constancy of the speed of light Eq.(2-8).

For instance, consider on the light sphere in K an arbitrary point defined by the coordinates

$$x = \alpha ct, \; y = \eta ct, \text{and} \; z = \lambda ct,$$

satisfying Eq. (2-6) (i.e., $\alpha^2 + \eta^2 + \lambda^2 = 1$). As a consequence of the Lorentz transformation, the x-coordinate expression $x = \alpha ct$ can be written—by substituting $x(x',t')$ and $t(x',t')$ from the Lorentz transformation Eqs. (2-2)—as

$$\gamma\left(x' + vt'\right) = \alpha c\gamma\left(t' + \frac{vx'}{c^2}\right),$$

simplified to

$$x'\left(1 - \alpha\frac{v}{c}\right) = \alpha ct'\left(1 - \frac{1}{\alpha}\frac{v}{c}\right),$$

—which would reduce to $x' = ct'$ for $\alpha = 1$. Squaring both sides of the above equation, we get

$$x'^2 = \alpha^2 c^2 t'^2 \left(\frac{1 - \frac{1}{\alpha}v/c}{1 - \alpha v/c}\right)^2.$$

Since, according to Lorentz transformation Eqs. (2-3), $y' = y$ and $z' = z$, we get

$$x'^2 + y'^2 + z'^2 = \alpha^2 c^2 t'^2 \left(\frac{1 - \frac{1}{\alpha}v/c}{1 - \alpha v/c}\right)^2 + (\eta^2 + \lambda^2)c^2 t^2,$$

which would reduce to the light sphere Eq. (2-7) in K', if $v = 0$ (which implies from the Lorentz transformation that $\gamma = 1$, and $t = t'$), or $\alpha = 1$ and $\eta = \lambda = 0$, returning $x'^2 = c^2 t'^2$ (and $x^2 = c^2 t^2$).

Now, dividing Eq. (2-11) by Eq. (2-12) yields

$$\left(\frac{x}{x'}\right)^2 = \left(\frac{ct}{ct'}\right)^2,$$

or

$$\frac{x}{x'} = \pm \frac{ct}{ct'}. \tag{2-21}$$

Assuming, for the time being, that $c > v$ (this assumption will turn out to be essential), then x and x' will always have the same sign (positive or negative), whether the light ray was emitted at $t_o = t'_o = 0$ in the positive or negative x-direction, with respect to the overlying K and K'. It follows that

$$\frac{x}{x'} \geq 0,$$

and given that

$$\frac{ct}{ct'} \geq 0,$$

Eq. (2-21) becomes

$$\frac{x}{x'} = \frac{ct}{ct'}. \tag{2-22}$$

Hence, Eq. (2-22) combined with Eqs. (2-11) and (2-12), leads to

$$c = \frac{x}{t} = \frac{x'}{t'}. \tag{2-23}$$

Eq. (2-23) can also be readily obtained using Fig. 2.1a, leading to

$$\frac{x}{t} = \frac{x'/\gamma}{t'/\gamma} = \frac{x'}{t'},$$

or Fig. 2.1b;

$$\frac{x'}{t'} = \frac{x/\gamma}{t/\gamma} = \frac{x}{t},$$

along with Eq. (2-8). In fact, using

$$\frac{x^2}{x'^2} = \frac{t^2}{t'^2},$$

in the expression resulting from dividing Eq. (2-8) by x'^2, leads to

$$\frac{t^2}{t'^2} - 1 = \frac{c^2}{x'^2}(t^2 - t'^2),$$

which yields Eq. (2-12), and Eq. (2-11) will follow when Eq. (2-12) is substituted into Eq. (2-8). Hence, Eq. (2-23) can be readily deduced.

Consequently, it can be concluded that the constancy of the speed of light for the light ray propagation in the relatively moving reference frames can be expressed by Eq. (2-23).

Direct Inference

It will be first shown that the constancy of the speed of light postulate is certainly unviable for relatively moving inertial reference frames, without a space-time distorting transformation. In fact, assuming the space-time is preserved (i.e., cannot be modified), the coordinates x and x' would then be related by the following equation with respect to K, in accordance with the Galilean transformation;

$$x' = x - vt. \tag{2-24}$$

Whereas, with respect to K', the same coordinates would be related by the following equation

$$x = x' + vt'. \tag{2-25}$$

Substituting Eq. (2-24) into Eq. (2-25), we get

$$t = t'. \tag{2-26}$$

Dividing both sides of Eqs. (2-24) and (2-25) by c, and applying the speed of light constancy principle as determined above ($c = x / t = x' / t'$), the following expressions are obtained;

$$t' = t - \frac{vx}{c^2},$$ (2-27)

and

$$t = t' + \frac{vx'}{c^2}.$$ (2-28)

Substituting Eq. (2-27) in Eq. (2-28), we get

$$x = x',$$ (2-29)

and replacing Eq. (2-29) in Eqs. (2-24) and (2-25) leads to the conflicting result of $v = 0$ at any $t > 0$, with the spatial coordinates x and x' being allowed to acquire non-zero values, according to Eqs. (2-24)–(2-29).

It follows that the set of Eqs. (2-24), (2-25), (2-27) and (2-28)—which will be referred to as (S1)—resulting from the Galilean transformation applied under the principle of the constancy of the speed of light, leads to the only conflicting solution $v = 0$, $x = x'$, and $t = t'$, binding the two reference frames together, although the relative motion of the reference frames is set as the main condition under which the equation set (S1) is derived. Consequently, the light speed constancy principle is unviable, at least in the case of no space-time distorting transformation.

On the other hand, although the equation set (S1) requires the conflicting binding of the two reference frames, it leads to the constancy of the speed of light general criteria given by Eq. (2-8), implying that the frames-binding requirement of the equation set (S1) remains applicable to Eq. (2-8).

In fact, Eqs. (2-24) and (2-27) lead to

$$x'^2 = x^2 + v^2t^2 - 2xvt,$$

and

$$c^2t'^2 = c^2t^2 + \frac{v^2x^2}{c^2} - 2xvt.$$

Eliminating $2xvt$ from the above two equations yields

$$x^2 + v^2t^2 - x'^2 = c^2t^2 + \frac{v^2x^2}{c^2} - c^2t'^2. \qquad (2\text{-}30)$$

Similarly, Eqs. (2-25) and (2-28) can lead to

$$-x'^2 - v^2t'^2 + x^2 = -c^2t'^2 - \frac{v^2x'^2}{c^2} + c^2t^2. \quad (2\text{-}31)$$

Adding Eqs. (2-30) and (2-31)—obtained from the equation set (S1)—and rearranging and simplifying the terms, returns Eq. (2-8):

$$x^2 - x'^2 = c^2t^2 - c^2t'^2.$$

Indeed, the addition of Eqs. (2-30) and (2-31) results in the following expressions,

$$2\left(x^2 - x'^2\right) + v^2\left(t^2 - t'^2\right) = 2c^2\left(t^2 - t'^2\right) +$$
$$+\frac{v^2}{c^2}\left(x^2 - x'^2\right);$$

$$\left(x^2 - x'^2\right)\left(2 - \frac{v^2}{c^2}\right) = c^2\left(t^2 - t'^2\right)\left(2 - \frac{v^2}{c^2}\right),$$

yielding the speed of light constancy principle equation,

$$x^2 - x'^2 = c^2\left(t^2 - t'^2\right).$$

Lorenz Transformation Re-derivation

Assuming that the principle of the speed of light constancy must result in a space-time distorting transformation, a length conversion by a factor of β along the direction of motion is hypothesized; the longitudinal length in one frame is scaled by a factor of β with respect to the other frame. This length conversion can therefore be expressed with respect to K and K', respectively, as follows.

$$x = vt + \beta x', \tag{2-32}$$

and

$$x' = -vt' + \beta x, \qquad (2\text{-}33)$$

where β is a positive real number.

Rearranging Eqs. (2-32) and (2-33), we can write

$$x' = \frac{1}{\beta}\left(x - vt\right), \qquad (2\text{-}34)$$

and

$$x = \frac{1}{\beta}\left(x' + vt'\right). \qquad (2\text{-}35)$$

Dividing both sides of Eqs. (2-34) and (2-35) by c, and applying the speed of light constancy principle equation, as demonstrated above through Eqs. (2-6) to (2-23) and restated here below;

$$c = \frac{x}{t} = \frac{x'}{t'}, \qquad (2\text{-}36)$$

the following expressions are obtained.

$$t' = \frac{1}{\beta}\left(t - \frac{vx}{c^2}\right), \qquad (2\text{-}37)$$

and

$$t = \frac{1}{\beta}\left(t' + \frac{vx'}{c^2}\right). \qquad (2\text{-}38)$$

Solving Eqs. (2-34), (2-35), (2-37) and (2-38) for β results in

$$\beta = \sqrt{1 - \frac{v^2}{c^2}},$$

or

$$\frac{1}{\beta} = \frac{1}{\sqrt{1 - \frac{v^2}{c^2}}} = \gamma. \qquad (2\text{-}39)$$

In fact, for $x' = 0$, Eqs. (2-34) and (2-38) yield $x = vt$, and $t' = \beta t$, respectively, reducing Eq. (2-37) to

$$\beta t = \frac{1}{\beta}\left(t - \frac{v^2 t}{c^2}\right);$$

therefore

$$\beta = \sqrt{1 - \frac{v^2}{c^2}}.$$

Conversely, for $x = 0$, Eqs. (2-35) and (2-37) yield $x' = -vt$, and $t = \beta t'$, respectively, reducing Eq. (2-38) to

$$\beta t' = \frac{1}{\beta}\left(t' - \frac{v^2 t'}{c^2}\right);$$

hence

$$\beta = \sqrt{1 - \frac{v^2}{c^2}}.$$

Alternatively, substituting Eqs. (2-35) and (2-38) in Eq. (2-34) results in an expression that can be simplified and solved for β, yielding Eq. (2-39).

We note that Eq. (2-39) is valid for $c > v$ only, thus satisfying our assumption made above in connection with the set criteria of the speed of light constancy principle.

It follows that, since $\beta < 1$, the hypothesized length conversion is a length contraction, as inferred from Eqs. (2-32) and (2-33).

The obtained set of Eqs. (2-34), (2-35), (2-37), (2-38), and (2-39) are the Lorentz transformation Eqs. (2-1), (2-2), and (2-4), representing the space-time transformation resulting from the reduced constancy of the speed of light principle given by Eq. (2-36).

It is noted that the velocity transformation Eqs. (2-5) can be readily derived from the conflicting Eqs. (2-24), (2-25), (2-27) and (2-28)—by dividing Eq. (2-24) by Eq. (2-27), and Eq. (2-25) by Eq. (2-28) —and working back from Eqs. (2-5), Eqs. (2-24), (2-25), (2-27) and (2-28) can be deduced, which implies that the Lorentz velocity transformation equations are merely invalid velocity criteria of the speed of light constancy principle, and independent of any space-time distorting transformation.

Lorentz Transformation Contradictions

The fact that the constancy of the speed of light principle is manifested as $c = x / t = x' / t'$ (or $\Delta x / \Delta t = \Delta x' / \Delta t'$), as deduced from the Lorentz transformation, is sufficient to conclude the invalidity of the Lorentz transformation—explicitly and fully constructed in this paper from this fact—since it implies that the origin coordinates of the relatively moving reference frames are undetermined with respect to each other.

In fact, under the condition of relative motion between the reference frames, the equation $c = x / t = x' / t'$ leads to the Lorentz transformation equations, as demonstrated earlier in the paper. Whereas, we conclude from the Lorentz transformation Eqs. (2-1) and (2-2) that when $x' = 0$, t' would not necessarily be zero, and conversely, when $t' = 0$, x' would not necessarily be zero. The same is true for x and t. It follows that, for the space origin of $K'\left(0,0,0,t'\right)$ at time $t' \neq 0$, the corresponding K x- and t-coordinates shall satisfy the relation

$$\frac{x}{t} = \frac{x'}{t'}$$

that would yield

$$x = x'\left(\frac{t}{t'}\right) = 0,$$

if t was determined. But, $x = 0$ results in undetermined t :

$$t = t'\left(\frac{x}{x'}\right) = \frac{0}{0},$$

making the above x-equation undetermined as well, thus leading to the set of K origin coordinates

$$\left(x = \frac{0}{0}, 0, 0, t = \frac{0}{0}\right)$$

with undetermined x and t .

And, for the time origin of $K'\left(x', y', z', 0\right)$, with spatial coordinates $\neq 0$, the corresponding K x- and t-coordinates shall satisfy the relation

$$\frac{t}{x} = \frac{t'}{x'},$$

that would yield

$$t = t'\left(\frac{x}{x'}\right) = 0,$$

if x was determined. But, $t = 0$ results in undetermined x:

$$x = x'\left(\frac{t}{t'}\right) = \frac{0}{0},$$

making the above t-equation undetermined as well, thus leading to the set of K origin coordinates

$$\left(x = \frac{0}{0}, y = y', z = z', t = \frac{0}{0}\right)$$

with undetermined x and t.

Hence, the constancy of the speed of light principle shall not principally be applicable at the reference frame space and time origins, restricting the time coordinate, and the spatial coordinate along the relative motion direction, from acquiring zero value (other than the set zero value at the initial overlaid-frames instant). Consequently, the Lorentz transformation, implicitly incorporating Eq. (2-36) as demonstrated earlier, results in various conflicts and unresolved paradoxes.

For instance, substituting Eq. (2-37) into Eq.(2-38), returns

$$t = \gamma \left(\gamma \left(t - \frac{vx}{c^2} \right) + \frac{vx'}{c^2} \right). \qquad (2\text{-}40)$$

Eq. (2-40) is simplified in the following steps.

$$t = \gamma^2 t - \frac{\gamma^2 vx}{c^2} + \frac{\gamma vx'}{c^2},$$

or

$$t\left(\gamma^2 - 1 \right) = \frac{vx}{c^2} \left(\gamma^2 - \frac{\gamma x'}{x} \right).$$

Using Eq. (2-36) in the above equation, we get

$$t\left(\gamma^2 - 1 \right) = \frac{vx}{c^2} \left(\gamma^2 - \frac{\gamma t'}{t} \right). \qquad (2\text{-}41)$$

With respect to Eq. (2-37), for $t' = 0$, the transformed t-coordinate with respect to K is $t = vx / c^2$ (t is undetermined with respect to Eq. (2-36) when $t' = 0$, as shown earlier, except at the initial overlaid-frames instant, the values of t and t' are set to zero). Therefore, for $t \neq 0$, Eq. (2-41) reduces to

$$t\left(\gamma^2 - 1 \right) = t\gamma^2, \qquad (2\text{-}42)$$

yielding the contradiction,

$$\gamma^2 - 1 = \gamma^2, \quad \text{or} \quad 0 = 1,$$

which is interpreted as the consequence of violating the restriction imposed by the light speed constancy principle on the coordinates (in this case setting $t' = 0$, equivalent to $t = vx \,/\, c^2$).

It follows that the transformation of $t' = 0$ to $t = vx \,/\, c^2$, for $x \neq 0$, by Lorentz transformation Eq. (2-37), is invalid, since it leads to a contradiction when used in Eq. (2-41), resulting from Lorentz transformation equations, for $t \neq 0$ (i.e., beyond the initial overlaid-frames instant satisfying $t = 0$ for $t' = 0$)—Letting $t = 0$ would satisfy Eq. (2-42), but another contradiction would emerge; the reference frames would be locked in their initial overlaid position, and no relative motion would be allowed, since in this case the corresponding coordinate to $t' = 0$ would be $t = vx \,/\, c^2 = 0$, yielding $v = 0$, as we're addressing the transformation of $t' = 0$ to $t = vx \,/\, c^2$ for $x \neq 0$.

Similar contradiction is obtained by substituting Eq. (2-38) into Eq. (2-37), using Eq. (2-36) in the resulting equation, and applying Eq. (2-38) for $t = 0$ $(t' = -vx' \,/\, c^2)$.

Furthermore, substituting Eq. (2-34) into Eq. (2-35), yields

$$x = \gamma\left(\gamma\left(x - vt\right) + vt'\right);$$

$$x\left(\gamma^2 - 1\right) = \gamma v\left(\gamma t - t'\right);$$

$$x\left(\gamma^2 - 1\right) = \gamma v t\left(\gamma - \frac{t'}{t}\right). \qquad (2\text{-}43)$$

Using Eq. (2-36) in Eq. (2-43), we get

$$x\left(\gamma^2 - 1\right) = \gamma v t\left(\gamma - \frac{x'}{x}\right). \qquad (2\text{-}44)$$

With respect to Eq. (2-34), for $x' = 0$, (corresponding to K' origin), the transformed x-coordinate with respect to K is $x = vt$ (x is undetermined with respect to Eq. (2-36) when $x' = 0$, as shown earlier, except at the initial overlaid-frame position, where the corresponding value to $x' = 0$ is $x = 0$). Therefore, for $x \neq 0$, Eq. (2-44) reduces to the following contradiction—interpreted as the consequence of violating the coordinate value restriction ($x' = 0$, equivalent to $x = vt$) imposed by the speed of light invariance principle.

$$x\left(\gamma^2 - 1\right) = x\gamma^2; \qquad (2\text{-}45)$$

$$\gamma^2 - 1 = \gamma^2, \quad \text{or} \quad 0 = 1.$$

It follows that the transformation of the x'-coordinate of K' origin ($x' = 0$) to $x = vt$, at time $t > 0$, with respect to K by Lorentz transformation Eq. (2-34), is invalid, since it leads to a contradiction when used in Eq. (2-44), resulting from Lorentz transformation equations, for $x \neq 0$ (i.e., beyond the initial overlaid-frames position satisfying $x = 0$ for $x' = 0$)—Letting $x = 0$ would satisfy Eq. (2-45), but another contradiction would emerge; the reference frames would be locked in their initial overlaid position, and no relative motion would be allowed, since in this case the corresponding coordinate to $x' = 0$ would be $x = vt = 0$, yielding $v = 0$, as we're addressing the transformation of $x' = 0$ to $x = vt$ for $t > 0$.

Yet, this conflicting condition of setting the spatial coordinate in the primed reference frame to zero under the speed of light invariance principle constitutes a vital strategy in the Lorentz transformation derivation, and the interpretation of the time dilation, in the special relativity formulation.[1,5]

Similar contradiction would follow upon substituting Eq. (2-35) into Eq. (2-34), using Eq. (2-36), and applying Eq. (2-35) for $x = 0$, $x' = -vt'$.

It follows that, the Lorentz transformation arrived at under the principle of the constancy of the speed of light is deemed to be refuted. Consequently, the length contraction hypothesis originally introduced as an *ad hoc*[3] to resolve the null result of the Michelson-Morley experiment[2] (inconsistency between experiment and theory with respect to a light ray fixed-length round trip travel time in the earth travel direction compared to that in the respective transverse direction) cannot be appropriately reconciled by the light velocity relativity principle space-time transformation.

The obtained Lorentz transformation contradictions for the particular cases of converting each of the spatial—along the relative motion direction—and time coordinates having a zero value in one reference frame to its corresponding value in the other frame, imply the general unviability of the Lorentz transformation equations.

CONFLICT RATIONALIZATION

In the constancy of the speed of light principle equations[3]

$$x^2 + y^2 + z^2 = c^2 t^2,$$

and

$$x'^2 + y'^2 + z'^2 = c^2 t'^2,$$

imposed as the governing aspect describing the space-time, (x, y, z, t) and (x', y', z', t') represent the space-time coordinates of an arbitrary light ray position vector in the reference frames K and K', respectively. Therefore, for instance, assigning the entity $x = vt$ to the x-coordinate of the origin of the reference frame

K' (i.e., transforming $x' = 0$ to $x = vt$ using the Lorentz transformation $x'(x, t)$ equation) imposes a conflict with the light ray position vector x-coordinate, which is forced in this case to take the value of vt. Therefore, imposing the constancy of the speed of light equations on the space-time coordinate systems prohibits the system coordinates from taking other values than those associated with, and describing the light ray position vector. Hence, the coordinates of the origin of the moving frame are in conflict with the light ray position vector coordinates. Therefore, the Lorenz transformation equations

$$x' = \gamma\left(x - vt\right),$$

and

$$x = \gamma\left(x' + vt'\right),$$

return the equations $x = vt$ and $x' = -vt'$ for the origin of $K'(x' = 0)$ and $K(x = 0)$, respectively, which are in contradiction with the constancy of the speed of light equations in which x and x' represent the x- and x'-coordinates of the light ray position vector.

Indeed, this justifies the appearance of the identified contradictions upon using the Lorentz transformation equations under the particular condition of $x' = 0$ (or $x = 0$), for which $x = vt$ (or $x' = -vt'$), with the constancy of the speed of light condition.

For instance, this conflict becomes clear when Eq. (2-10) is inserted in Eq. (2-8) for the particular case of $x' = 0$ (i.e., $ct' = 0$) corresponding to $x = vt$, returning

$$v^2 t^2 = c^2 t^2, \text{ or } v = \pm c.$$

Furthermore, the same conflict is revealed when Eqs. (2-9) and (2-10) are inserted in Lorentz transformation time Eqs. (2-1) and (2-2) for the particular cases of $t' = 0$ and $t = 0$, respectively, yielding

$$t = \frac{vx}{c^2} = \frac{vct}{c^2}, \text{ or } v = c;$$

$$t' = \frac{-vx'}{c^2} = \frac{-vct'}{c^2}, \text{ or } v = -c.$$

APPARENT SPACE-TIME TRANSFORMATION

Let's consider the classical coordinate transformation Eq. (2-24), and hypothesize a general length conversion factor β from the perspective of the reference frame K, under the light speed constancy assumption:

$$x = vt + \beta x',$$

or

$$x' = \frac{1}{\beta}\left(x - vt\right). \qquad (2\text{-}46)$$

Dividing Eq. (2-46) by c and applying the constancy of the speed of light Eq. (2-36), we get

$$t' = \frac{1}{\beta}\left(t - \frac{vx}{c^2}\right). \qquad (2\text{-}47)$$

Substituting $x = vt + \beta x'$ from Eq. (2-46) into Eq. (2-47), the following operations are performed to solve for t.

$$t' = \frac{1}{\beta}\left(t - \frac{v\left(vt + \beta x'\right)}{c^2}\right),$$

or

$$t' = \frac{t}{\beta} - \frac{v^2 t}{\beta c^2} - \frac{vx'}{c^2},$$

then

$$\frac{t}{\beta}\left(1 - \frac{v^2}{c^2}\right) = t' + \frac{vx'}{c^2}.$$

Letting

$$\gamma = \frac{1}{\sqrt{1 - \dfrac{v^2}{c^2}}},$$

we get

$$\frac{t}{\beta}\left(\frac{1}{\gamma^2}\right) = t' + \frac{vx'}{c^2},$$

or

$$t = \beta\gamma^2\left(t' + \frac{vx'}{c^2}\right). \tag{2-48}$$

Substituting Eq. (2-48) into Eq. (2-46), rearranging, and simplifying the terms, we get

$$x = \beta\gamma^2\left(x' + vt'\right). \tag{2-49}$$

Indeed,

$$x = vt + \beta x';$$

$$x = v\beta\gamma^2\left(t' + \frac{vx'}{c^2}\right) + \beta x';$$

$$x = v\beta\gamma^2 t' + \beta x'\gamma^2\left(\frac{1}{\gamma^2} + \frac{v^2}{c^2}\right).$$

Since the latter equation terms between the brackets add to unity, it reduces to Eq. (2-49).

For the particular case of $\beta = 1 / \gamma$, Eqs. (2-46) to (2-49) take the form of the known Lorentz transformation equations. In addition, the relativistic velocity transformation equations can be derived from Eqs. (2-46) to (2-49), irrespective of the value of β.

It can be concluded from the transformation resulting from the light velocity invariance principle, that for a length factor of β with respect to K, Eqs. (2-46) and (2-47) lead to the following equations for simultaneous $(\Delta t = 0)$ and co-local $(\Delta x = 0)$ events, respectively;

$$x = \beta x', \tag{2-50}$$

and

$$t = \beta t'. \tag{2-51}$$

Whereas, for simultaneous $(\Delta t' = 0)$ and co-local $(\Delta x' = 0)$ events with respect to K', we have, respectively, from Eqs. (2-49) and (2-48),

$$x' = \frac{1}{\beta}\left(\frac{1}{\gamma^2}\right)x, \tag{2-52}$$

and

$$t' = \frac{1}{\beta}\left(\frac{1}{\gamma^2}\right)t. \tag{2-53}$$

For the case of preserved space-time where the length factor is $\beta = 1$, if we consider the events of a light ray being emitted and returned, after being reflected, to the same point ($\Delta x' = 0$) in the longitudinal direction in K', it can be easily shown that, in line with the constancy of the light speed, the light ray travel time in K would be

$$T = \gamma^2 \left(\frac{2L}{c} \right), \qquad (2\text{-}54)$$

where $2L$ is the round trip length. According to Eq. (2-52), the corresponding round trip length in K' would be

$$2L' = \frac{2L}{\gamma^2}, \qquad (2\text{-}55)$$

and the corresponding travel time in K' becomes, according to Eq. (2-53),

$$T' = \frac{T}{\gamma^2} = \frac{\gamma^2 \left(2L / c \right)}{\gamma^2} = \frac{2L}{c}. \qquad (2\text{-}56)$$

Now, with the length factor of β being applied in K, the round trip length becomes $\beta(2L)$, and the light ray round trip travel time in K becomes, using Eq. (2-54),

$$T = \gamma^2 \left[\frac{\beta(2L)}{c} \right] = \beta\gamma^2 \left(\frac{2L}{c} \right), \qquad (2\text{-}57)$$

whereas, according to Eq. (2-53), the corresponding travel time in K', using Eq. (2-57), becomes

$$T' = \frac{1}{\beta} \left(\frac{T}{\gamma^2} \right) = \frac{1}{\beta} \left(\frac{\beta\gamma^2 \left(2L / c \right)}{\gamma^2} \right) = \frac{2L}{c}, \qquad (2\text{-}58)$$

while, from Eq. (2-52) the corresponding round trip length becomes

$$2L' = \frac{1}{\beta} \left(\frac{\beta(2L)}{\gamma^2} \right) = \frac{2L}{\gamma^2}. \qquad (2\text{-}59)$$

It follows from Eqs. (2-55), (2-56), (2-58) and (2-59) that the transformed longitudinal light ray round trip length and travel time in K' are independent of the introduced length conversion factor β in K , and always converted to $2L / \gamma^2$ and $2L / c$, respectively.

It becomes then obvious that the transformation (converting from K coordinates to K' coordinates) resulting from a length conversion factor with respect to K under the application of the contradictory light velocity invariance criteria $(c = x / t = x' / t')$, simply reverses the length factor to recover the original length in K, and scales the recovered length down by a factor of $1/\gamma^2$ so that the local time in K' is obtained. This is indeed an amazing, tricky transformation; when the length conversion factor is $1/\gamma$, thus changing the travel time in K from $\gamma^2\left(2L/c\right)$ to $\gamma\left(2L/c\right)$, the light velocity constancy resulting transformation would reverse the length factor, returning the original time of $\gamma^2\left(2L/c\right)$ in K, and apply a new length factor of $1/\gamma^2$ converting the travel time to $2L/c$ in K', with a net length factor of $\left(1/\gamma\right)^{-1}\left(1/\gamma^2\right)=1/\gamma$, giving the impression of a space-time distorting transformation, with a time dilation factor of γ and a length contraction of $1/\gamma$, although the actual length contraction factor in K' is $1/\gamma^2$!

It follows that the length conversion factor of $1/\gamma$ is nothing but a particular factor resulting in [conflicting] symmetrical transformation equations, when applying the restricted speed of light constancy principle on the classical spatial transformation equation with a length conversion factor. Otherwise, any length conversion factor β introduced to the classical spatial coordinate equation $x=vt+x'$ (changing it to $x=vt+\beta x'$) under the restricted assumption of the constancy of the speed of light, would result in inapplicable time and space transformation Eqs.—(2-46) to (2-49)— invariantly satisfying the basic criteria of the light velocity assumption given by Eq. (2-8).

Indeed, squaring both Eqs. (2-46) and [$c \times$ Eq. (2-47)], and eliminating the similar term from the resulting two equations, leads to

$$x^2 + v^2 t^2 - x'^2 \beta^2 = c^2 t^2 + \frac{v^2 x^2}{c^2} - c^2 t'^2 \beta^2. \quad (2\text{-}60)$$

Similar application of Eqs. (2-49) and (2-48) will result in

$$-x'^2 - v^2 t'^2 + \frac{x^2}{\beta^2 \gamma^4} = -c^2 t'^2 - \frac{v^2 x'^2}{c^2} + \frac{c^2 t^2}{\beta^2 \gamma^4}. \quad (2\text{-}61)$$

—It should be noted that Eqs. (2-60) and (2-61) reconfirm the equalities $x^2 = c^2 t^2$, and $x'^2 = c^2 t'^2$.

Adding Eqs. (2-60) and (2-61), and simplifying and rearranging the terms, leads to

$$x^2 - x'^2 = c^2 t^2 - c^2 t'^2 + \frac{1}{\beta^2 \gamma^2} \left(c^2 t^2 - x^2 \right) - $$
$$-\beta^2 \gamma^2 \left(c^2 t'^2 - x'^2 \right),$$

which reduces to the constancy of the speed of light Eq. (2-8):

$$x^2 - x'^2 = c^2t^2 - c^2t'^2.$$

Finally, the above discussion, carried out from Eq. (2-24), from the perspective of K, can also be repeated based on Eq. (2-25), from the perspective of K', with identical results being obtained.

CONCLUSION

Analysis of the Lorentz transformation revealed mathematical restrictions in terms of the deduced, simplified form of the constancy of the speed of light equations residing in the transformation. The Lorentz transformation, readily reconstructed using these basic, restricted light velocity invariance equations, resulted in mathematical contradictions. The principle of the constancy of the speed of light was thus demonstrated to be an unviable assumption, and the ensuing Lorentz transformation was subject to refutation. Rationalization of the revealed contradictions was established. The actual interpretation of the Lorentz transformation demonstrated the unreal aspect of the space-time conversion attributed to the transformation.

Chapter 3

ROOT CAUSE OF THE PROBLEMS WITH LORENTZ TRANSFORMATION

In this chapter the Lorentz Transformation is shown to be merely a set of restricted equations stemmed from the Galileo transformation applied to a particular conversion reflecting the theorized principle of the speed of light invariance implemented in the direction of the relative motion between the inertial reference frames. Consequently, the Lorentz transformation is shown to be restricted to time and longitudinal space coordinates different from zero. The deduction of the time dilation and length contraction becomes unfeasible under such restrictions. It follows that the Lorentz transformation possesses no other effects than mathematically expressing the speed of light postulate in the relative motion direction. In addition, the application of the Lorentz transformation to events having restricted coordinates is shown to result in mathematical contradictions. Furthermore, the particular terms, erroneously generalized in the Lorentz transformation, are replaced with their unaltered expressions, resulting in a transformation conforming to the speed of light postulate, but having detrimental consequences on the Special Relativity predictions. The essential anomaly in the Lorentz time transformation equations leading to their contradictions is identified, and the Special Relativity "established" predictions turn out to be overwhelmingly refuted.

BACKGROUND

The Lorentz transformation (LT) equations constitute the backbone of the Special Relativity (SR) theory in which their interpretations lead to the peculiar predictions of the space-time distortion characterized by the length contraction and time dilation. The SR predictions have led to numerous paradoxes, consistently generating critical publications on the SR validity,[6,8,10,12,14] particularly the clock paradox expressed in what's become known as the twin paradox, discussed in details in a critical study[14] challenging the viability of the SR. Contradictions resulting from the relativistic length contraction have been revealed in a study on the inconsistency of the relativity principle with the SR predictions.[15]

It has been claimed that the relativistic time dilation has been verified experimentally, the Hafele-Keating experiment[16] having the most celebrity. However, their experimental results are questionable for many reasons.[12,14] Another prominent experiment is the high energy particles lifetime measurement,[17,18] which is also shown to be unreliable for various causes.[12,14]

The objective of this chapter is to scrutinize the formulation of the LT equations as to identify the anomalies leading to their persisting paradoxical outcomes.

The LT in SR is derived on the basis of the relativity principle and the constancy of the speed of light postulate.[1,5,19] The sought transformation, converting between the space and time coordinates of two inertial reference frames, say $K(x,y,z,t)$ and $K'(x',y',z',t')$, in relative motion at speed v, was assumed to take the following general form

$$x' = ax + bt$$
$$y' = y; z' = z$$
$$t' = kx + mt$$

where a, b, k, and m are unknown real terms.

Whereas, the constancy of the speed of light postulate was expressed by the assumption that a spherical light wave front, emitted from the coinciding inertial frame origins, would be observed as a light sphere centered at the frame origin, with its radius being expanded at the speed of light c, with respect to either frame:

$$x^2 + y^2 + z^2 = c^2 t^2$$
$$x'^2 + y'^2 + z'^2 = c^2 t'^2$$

leading to

$$x^2 - x'^2 = c^2t^2 - c^2t'^2.$$

In the customary derivation of the Lorentz transformation, the above proposed space and time transformation equations along with the latter speed of light constancy equation—applied with some given particular conditions and using the transformation symmetry assumption—would be solved for the unknown terms, yielding the following LT equations:

$$x' = \gamma\left(x - vt\right)$$
$$y' = y, \ \ z' = z$$
$$t' = \gamma\left(t - \frac{vx}{c^2}\right)$$
$$\gamma = \frac{1}{\sqrt{1 - \dfrac{v^2}{c^2}}}$$

The above approach is rather complex, which makes inconsistent operations performed in the derivation process easily bypassed. For instance, the above constancy of the speed of light equation was obtained in a published work[19] on SR through constructing it from the basic conversion expressions, $x = ct$; $x' = ct'$, presenting the speed of light invariance in the relative motion direction:

$$x^2 = c^2 t^2; \ x^2 - c^2 t^2 = 0$$
$$x'^2 = c^2 t'^2; \ x'^2 - c^2 t'^2 = 0$$
$$x^2 - x'^2 = c^2 t^2 - c^2 t'^2$$

Obviously, the intrinsic property of the basic expressions $x = ct$; $x' = ct'$, requiring $x = 0$ when $t = 0$ —thus leading to $x' = 0$ and $t' = 0$— is lost in the above constructed speed of light equation. To remedy this inconsistency, the above constructed equation should be restricted to non-zero coordinate values.

Consequently, to avoid the encountered inconsistencies in the above conventional derivation approach, a straight forward method is used in this study to derive and reveal the innate limitations of the Lorentz transformation.

The speed of light constancy principle equations, as well as the Lorentz transformation, have been the subject of an analytical study,[20] in which mathematical contradictory results, attributed to the LT and the speed of light postulate, have been unveiled. This study provides supplementary materials to the said work, in which the attained conclusions are reconfirmed by addressing the LT from a different perspective using a direct derivation approach—rather than working backward through analyzing the given Lorentz transformation—leading to the same detrimental contradictions. In addition, this study takes a step further to correct the contradictory terms in the Lorentz transformation, resulting in the actual transformation that should follow from the SR constancy of the speed of light postulate. The obtained transformation effect is in total contradiction with the SR essential predictions.

LIMITATIONS OF THE LORENTZ TRANSFORMATION

Consider two inertial reference frames, $K(x,y,z,t)$ and $K'(x',y',z',t')$, in relative uniform motion along the overlapped x- and x'-axes, at a speed v. The transformation relating the space and time coordinates of the two frames is to be determined. If the space dimension was considered to be unscaled from one frame to another, the coordinate conversion equation would then be governed by the Galilean transformation, namely

$$x' = x - vt \qquad (3\text{-}1)$$

with unchanged y and z coordinates (i.e., $y = y'$; $z = z'$).

For the generalized case where the space coordinate—in the relative motion direction—is assumed to be deformed, while maintaining the linearity of the transformation, introducing coordinate conversion factors would then be hypothesized,

inferring that the general spatial transformation shall
have the following linear form;

$$x' = \gamma x + \beta t, \qquad (3\text{-}2)$$

where γ and β are real terms to be determined—y
and z remain invariant.

For both cases described by Eqs. (3-1) and (3-2),
the origin of K' is traveling at speed v with respect
to K origin. Therefore, we can conclude that the
coordinate $x' = 0$ in K' would be transformed to
$x = vt$ in K, by both equations. Hence, plugging the
particular conversion $x' = 0$; $x = vt$ in the general
transformation Eq. (3-2) yields the particular equation
$0 = \gamma vt + \beta t$, or $\beta = -\gamma v$ (for $t \neq 0$), leading to a
simplified general transformation equation

$$x' = \gamma \left(x - vt \right). \qquad (3\text{-}3)$$

Furthermore, under the principle of the constancy
of the speed of light, another particular conversion
related to the x-coordinate of the tip point of a light
ray propagating in the relative motion direction is
readily available, and can be expressed as
$x = ct$; $x' = ct'$, which, when plugged in Eq. (3-3),
leads to the particular equation

$$ct' = \gamma\left(ct - \frac{vx}{c}\right);$$

$$t' = \gamma\left(t - \frac{vx}{c^2}\right), \qquad (3\text{-}4)$$

with the above restriction $t \neq 0$ being maintained, leading to the additional restriction of $x \neq 0$, since $t = x/c$ is used to get the expression vx/c^2 in Eq. (3-4).

Now, owing to the fact that the reference frame K is traveling at a speed of $-v$ with respect to K', and to the essential symmetrical property of the transformation with respect to the reference frames, the inverse of the general transformation given by Eq. (3-3) can be written as

$$x = \gamma\left(x' + vt'\right), \qquad (3\text{-}5)$$

which must be as well restricted—by symmetry—to $t' \neq 0$.

Similarly, under the principle of the constancy of the speed of light, plugging the particular conversion of the tip point x'-coordinate of a light ray propagating in the relative motion direction, expressed as $x' = ct'$; $x = ct$, in the general transformation Eq. (3-5) leads to the particular equation

$$t = \gamma \left(t' + \frac{vx'}{c^2} \right),$$ (3-6)

equally maintaining the above restriction $t' \neq 0$, leading to $x' \neq 0$.

Substituting Eqs. (3-3) and (3-4) in Eq. (3-6), leads after simplification to

$$\gamma = \frac{1}{\sqrt{1 - \frac{v^2}{c^2}}}.$$ (3-7)

It follows that Eqs. (3-3)–(3-7) constitute the Lorentz transformation—and its inverse—although Eqs. (3-4) and (3-6) are shown to be merely particular equations obtained from the special conversion $x = ct;\ x' = ct'$, expressing the constancy of the speed of light principle in the direction of the relative motion, when plugged in the general transformation Eqs. (3-3) and (3-5). In addition, as demonstrated above, the LT Eqs. (3-3)–(3-7) are restricted to values of x, t, x', and t' different from zero.

Furthermore, since γ was determined with the use of the particular Eqs. (3-4) and (3-6), Eqs. (3-3) and (3-5) would bear the same limitations as Eqs. (3-4) and (3-6). It follows that all LT equations are limited to the particular conversion $x = ct;\ x' = ct'$, and to coordinate values not equal to zero. These results have been confirmed in an earlier study through mathematical analyses of the Lorentz transformation.[13]

CONFLICTING FINDINGS

The invalid generalization of the particular Eqs. (3-4) and (3-6) would result in mathematical conflicts. Indeed, substituting Eq. (3-4) into Eq. (3-6), returns

$$t = \gamma\left(\gamma\left(t - \frac{vx}{c^2}\right) + \frac{vx'}{c^2}\right),$$

which can be simplified to

$$t\left(\gamma^2 - 1\right) = \frac{vx}{c^2}\left(\gamma^2 - \frac{\gamma x'}{x}\right). \qquad (3\text{-}8)$$

Since, as shown earlier, Eqs. (3-4) and (3-6) already satisfy the conversion $x = ct$; $x' = ct'$, then Eq. (3-8) can be written as

$$t\left(\gamma^2 - 1\right) = \frac{vx}{c^2}\left(\gamma^2 - \frac{\gamma t'}{t}\right). \qquad (3\text{-}9)$$

If Eqs. (3-4), (3-6) and (3-9) were generalized (i.e., applied to conversions other than $x = ct$; $x' = ct'$, or

$t = x / c; t' = x' / c$), and particularly applied to an event with the restricted time $t' = 0$, then according to Eq. (3-4), the transformed t-coordinate with respect to K would be $t = vx / c^2$. Consequently, for $t \neq 0$, Eq. (3-9) would reduce to

$$t\left(\gamma^2 - 1\right) = t\gamma^2, \qquad (3\text{-}10)$$

yielding the contradiction,

$$\gamma^2 - 1 = \gamma^2, \quad \text{or} \quad 0 = 1.$$

It follows that the conversion of the restricted time coordinate $t' = 0$ to $t = vx / c^2$, for $x \neq 0$, by LT Eq. (3-4), is proved to be invalid, since it leads to a contradiction when used in Eq. (3-9), resulting from the LT equations for $t \neq 0$ (i.e., beyond the initial overlaid-frames instant satisfying $t = 0$ for $t' = 0$) —Letting $t = 0$ would satisfy Eq. (3-10), but another contradiction would emerge; the reference frames would be locked in their initial overlaid position, and no relative motion would be allowed, since in this case the corresponding coordinate to $t' = 0$ would be $t = vx / c^2 = 0$, yielding $v = 0$, as we're addressing the conversion of $t' = 0$ to $t = vx / c^2$ for $x \neq 0$.

A similar contradiction is obtained by substituting Eq. (3-6) into Eq. (3-4), and applying Eq. (3-6) for the

conversion $t = 0$; $t' = -vx' / c^2$ of the restricted time
coordinate $t = 0$.

Furthermore, substituting Eq. (3-3) into Eq. (3-5),
yields

$$x = \gamma \left(\gamma \left(x - vt \right) + vt' \right);$$

$$x \left(\gamma^2 - 1 \right) = \gamma v \left(\gamma t - t' \right);$$

$$x \left(\gamma^2 - 1 \right) = \gamma v t \left(\gamma - \frac{t'}{t} \right). \qquad (3\text{-}11)$$

Since Eqs. (3-3) and (3-5), along with Eqs. (3-4)
and   (3-6),   already   satisfy   the   conversion
$x = ct$; $x' = ct'$, Eq. (3-11) can be written as

$$x \left(\gamma^2 - 1 \right) = \gamma v t \left(\gamma - \frac{x'}{x} \right). \qquad (3\text{-}12)$$

If Eqs. (3-3), (3-5) and (3-12) were generalized
(i.e.,   applied   to   conversions   other   than
$x = ct$; $x' = ct'$), and particularly applied to an event
with the restricted coordinate $x' = 0$, then according
to Eq. (3-3), the transformed x-coordinate with
respect to K would be $x = vt$, Consequently, for
$x \neq 0$, Eq. (3-12) would reduce to

$$x \left(\gamma^2 - 1 \right) = x\gamma^2; \qquad (3\text{-}13)$$

$$\gamma^2 - 1 = \gamma^2, \quad \text{or} \quad 0 = 1.$$

It follows that the conversion of the restricted space coordinate $x' = 0$ of K' origin to $x = vt$, at time $t > 0$, with respect to K by LT Eq. (3-3), is invalid, since it leads to a contradiction when used in Eq. (3-12) resulting from LT equations, for $x \neq 0$ (i.e., beyond the initial overlaid-frames position satisfying $x = 0$ for $x' = 0$)—Letting $x = 0$ would satisfy Eq. (3-13), but another contradiction would emerge; the reference frames would be locked in their initial overlaid position, and no relative motion would be allowed, since in this case the corresponding coordinate to $x' = 0$ would be $x = vt = 0$, yielding $v = 0$, as we're addressing the conversion of $x' = 0$ to $x = vt$ for $t > 0$.

A similar contradiction would follow upon substituting Eq. (3-5) into Eq. (3-3), and applying Eq. (3-5) for the conversion $x = 0$; $x' = -vt'$ of the restricted space coordinate $x = 0$.

It is worth mentioning that another conflict would emerge upon letting $t' = 0$ in the conversion $x' = ct'$; $x = ct$, since this results in $x' = 0$ (a restricted coordinate for Lorentz transformation) which is converted into $x = vt$ by Eq. (3-3), leading to the conflicting equality $vt = ct$, or $v = c$. — Alternatively, $t' = 0$ is restrictively converted into

$t = vx / c^2$ by Eq. (3-4), leading to $x = vx / c$, or $v = c$.

It follows that, the application of the LT to events having any of the determined restricted coordinates (i.e., x, x', t, or t' is equal to zero) is unfeasible, as it leads to contradictions. Consequently, the interpretation of the time dilation and length contraction would not be possible, since the former requires co-local events (i.e., $x' = 0$) and the latter simultaneous events (i.e., $t = 0$).

THE ACTUAL "LIGHT SPEED POSTULATE" TRANSFORMATION

Going back to our LT derivation section above, let's apply the general transformation Eq. (3-3) to the particular conversion of $x = ct$; $x' = ct'$, leading us to the time transformation equation

$$ct' = \gamma \left(ct - vt \right);$$

$$t' = \gamma \left(t - \frac{vt}{c} \right),$$

without effecting the replacement of t with x / c in the term vt / c—that would change the above equation to LT Eq. (3-4), shown to be inconsistent when applied to events with zero time (t or t') or zero longitudinal spatial coordinate (x or x')—ending up with the following time transformation equation

$$t' = \gamma t \left(1 - \frac{v}{c}\right), \qquad (3\text{-}14)$$

which is a general time equation since it only involves the time variables.

Similarly, when applied to the particular conversion of $x = ct$; $x' = ct'$, the general transformation Eq. (3-5) can lead to

$$t = \gamma t' \left(1 + \frac{v}{c}\right). \qquad (3\text{-}15)$$

Substituting Eq. (3-14) into Eq. (3-15) leads after simple simplification to

$$\gamma = \frac{1}{\sqrt{1 - \frac{v^2}{c^2}}}. \qquad (3\text{-}16)$$

It is worth mentioning that dividing Eq. (3-3) by Eq. (3-14) leads to the following velocity transformation

$$u' = \frac{x'}{t'} = \frac{\gamma\left(x - vt\right)}{\gamma t\left(1 - \dfrac{v}{c}\right)};$$

$$u' = \frac{u - v}{1 - \dfrac{v}{c}}.$$

Similarly, dividing Eq. (3-5) by Eq. (3-15) leads to

$$u = \frac{u' + v}{1 + \dfrac{v}{c}},$$

where u and u' are the velocity of an object with respect to K and K', respectively, traveling in the relative motion direction.

It is noted that the obtained velocity transformation returns the speed of light c when either u or u' is replaced with c.

THE CONTRADICTION ROOT CAUSE

It follows that the light speed constancy principle leads to Eqs. (3-14) and (3-15) presenting the relationship between the time coordinates t and t' in the two reference frames K and K'. However, these

equations are in total disagreement with the respective
LT equations given by

$$t' = \gamma\left(t - \frac{vx}{c^2}\right) \tag{3-17}$$

$$t = \gamma\left(t' + \frac{vx'}{c^2}\right) \tag{3-18}$$

To identify the cause of this discrepancy, let's
rewrite Eqs. (3-14) and (3-15) in the following form

$$t' = \gamma\left(t - \frac{vt}{c}\right) \tag{3-19}$$

$$t = \gamma\left(t' + \frac{vt'}{c}\right) \tag{3-20}$$

Obviously, the term $vt\,/\,c$ in Eq. (3-19) represents
the time it takes a light signal to propagate across the
distance vt traveled by K' origin with respect to K
at time t. Now, if, for instance, two light signals,
separated by a time interval t, were emitted from the
origin of K' (i.e, $x' = 0$), the distance vt becomes
equal to x, and the term $vt\,/\,c$ reduces to $x\,/\,c$,
simplifying Eq. (3-19) to

$$t' = \gamma\left(t - \frac{x}{c}\right).$$

It should be noted that x / c in the above particular equation (for the particular case of $x' = 0$) is the light signal propagation time through the distance x (that replaced the general term vt for the special case of $x' = 0$), and not the time t (equals to the travel time of the reference frame K' through the distance x at the speed v). Also, setting $x = 0$ in the above equation doesn't merely mean that $t' = \gamma t$, since the above equation is resulting from the time transformation Eq. (3-19) for $x = vt$, i.e. for $x' = 0$, and the condition of having both x and x' equal to zero corresponds to $t = t' = 0$, or to no relative motion between the reference frames, i.e. to $v = 0$, equivalent to $\gamma = 1$.

Comparing the above equation with LT Eq. (3-17), we notice the contradiction in the term vx / c^2, requiring $v = c$.

The same reasoning can be applied to Eqs. (3-18) and (3-20) to draw a similar contradiction from Eq. (3-18) (i.e., $v = -c$).

Even more, by a simple analysis of the spatial LT equation using the light speed constancy, it is obvious that, with respect to K, the time t' it takes a light signal, emitted from the point of the coinciding origins at $t = t' = 0$ to travel a distance x' in K' is equal to the time t for the signal to travel the corresponding distance x in K less the signal travel time of the distance vt travelled by the origin of K' at the time t, corrected by the relativistic factor γ. In other words, an event occurring in K' [origin] at the time t with respect to K has already occurred with respect to K' at the time t' equal to t less the signal time of travel from the position of K' [origin] at the time t to K origin, corrected by the relativistic factor γ.

Therefore, the term vx / c^2 in the Lorentz time transformation Eq. (3-17) must be the [uncorrected] time it takes the light signal to travel the distance between the origins at the time t with respect to K, or

$$\frac{vx}{c^2} = \frac{vt}{c},$$

leading to

$$x = ct.$$

Hence, the term x in the LT Eq. (3-17) is actually confined to the value of ct (erroneously replaced with x), which is contradicted with the fact that x takes the value of vt when $x' = 0$, making $v = c$ for $x' = 0$; this is indeed the source of the LT contradiction obtained earlier for $x' = 0$ (as well as when $x = 0$). A similar contradiction emerging from LT Eq. (3-18) can be demonstrated (i.e., erroneously using $x' = ct'$ in the term vt'/ c leads to $v = -c$, since for $x = 0$; $x' = -vt'$).

Furthermore, when $t' = 0$, LT Eq. (3-17) leads to $t = vx / c^2$. But, as shown above, $x = ct$ in Eq.

(3-17), yielding the contradiction $t = vct / c^2$, or $v = c$.

Similarly, LT Eq. (3-18) can lead to a similar contradiction for $t = 0$ (i.e., $v = -c$).

This confinement of the x and x' coordinates to $x = ct$ and $x' = ct'$ in the LT equations can also be demonstrated by replacing Eq. (3-15) in Eq. (3-3), and Eq. (3-14) in Eq. (3-5), returning, respectively

$$x = \gamma\left[x'\left(\frac{1}{\gamma^2} + \frac{v^2 t'}{cx'}\right) + vt'\right],$$

and

$$x' = \gamma\left[x\left(\frac{1}{\gamma^2} + \frac{v^2 t}{cx}\right) - vt\right],$$

requiring $x' = ct'$ and $x = ct$, to yield the LT Eqs. (3-5) and (3-3), respectively, as well as Eqs. (3-17) and (3-18) when this requirement is applied to Eqs. (3-19) and (3-20).

It follows that the Lorentz time transformation Eqs. (3-17) and (3-18) are invalid, and the unaltered structure of these equations is given by Eqs. (3-14) and (3-15), which totally overthrow all of the SR predictions in terms of the length contraction and time dilation, and their resulting interpretations. Moreover, the actual transformation emerging from the light speed postulate has no realistic interpretation, ending up with an unrealistic postulate.

CONCLUSION

The LT is demonstrated to be restricted to events having non-zero time coordinates and non-zero space coordinates along the reference frames axes parallel to the relative motion direction. With such imposed coordinate restrictions, the effects of the time dilation and length contraction become unfeasible. Furthermore, the LT is shown to be limited to merely expressing mathematically the speed of light postulate in the relative motion direction, with no practical results or predictions being obtained from its application.

In addition, The Lorentz time transformation equations are demonstrated to erroneously confine the involved spatial coordinates (in the terms $vx \, / \, c^2$ and $vx' \, / \, c^2$) to the specific values of $x = ct$ and $x' = ct'$, which results in conflict with the frame origin coordinates with respect to one another (i.e., $x = vt$ and $x' = -vt'$). In using the correct term in these equations, the resulting transformation is found to be in total disagreement with the Lorentz transformation, leading to the refutation of its predictions.

Chapter 4

INCOMPATIBILITY OF THE LIGHT SPEED POSTULATE WITH THE COORDINATE TRANSFORMATION SYMMETRY ASSUMPTION

The speed of light postulate is closely examined from the perspective of two inertial reference frames—unprimed ('stationary') and primed ('traveling')—in relative motion, revealing that the speed of light postulate actually requires length contraction with respect to the unprimed reference frame, and length expansion with respect to the primed frame. It is shown that when symmetry is imposed on the inverse length transformation (i.e., to make it exhibit the same length contraction from the perspective of the primed frame), the common length contraction factor becomes nothing but the Lorentz contraction factor γ. However, this would necessarily result in $\gamma = 1$, implying that the frames are being at rest with respect to each other, and thus refuting the special relativity predictions! When the coordinate's transformation symmetry assumption is applied on the direct transformation resulting from the light speed postulate,—which is shown incompatible with this assumption—the Lorentz transformation and its inverse are erroneously obtained; it is shown to be restricted to certain coordinate relations, resulted in mathematical contradictions, and thus demonstrated to be unviable.

BACKGROUND

The Lorentz transformation, providing interrelation between the coordinates of two inertial reference frames in relative motion, forms the heart of the Special Relativity Theory. Einstein[1,5] mainly derived the transformation on the basis of two principles: 1- the principle of relativity, stating that the laws of physics are the same in all inertial reference frames, and 2- the speed of light principle, postulating that the speed of light in vacuum is invariant with respect to all inertial frames of reference.

Yet, another essential tool used in the Lorentz transformation derivation is that the direct and inverse transformations exhibit mutually symmetrical property; that is, the inverse transformation equation can be deduced from the direct one by swapping the coordinates and reversing the velocity sign. This is essentially the result of the isotropic property of space, combined with the first principle of the special relativity. This assumption is rather intuitive. However,

Incompatibility of the Light
Speed Postulate with the Coordinate
Transformation Symmetry Assumption   93

in this Chapter, it is demonstrated that the speed of light principle deviates from this "law" of transformation symmetry. That is, the speed of light principle consequent direct transformation from the perspective of one frame is not symmetrical relative to the corresponding inverse transformation from the perspective of the other frame in relative translational motion with respect to the first frame. It is shown that this fact has a fatal outcome in regard to the coherence of the special relativity, in agreement with the findings of earlier studies.[20,21]

FATAL CONSEQUENCE OF THE LIGHT SPEED POSTULATE AND THE TRANSFORMATION SYMMETRY

Let $K(x, y, z, t)$ be a coordinate system attached to a reference frame K , and let $K'(x', y', z', t')$ be another coordinate system attached to a reference frame K' in relative translational motion at a uniform velocity v, with respect to K.

A light ray is emitted when the two frames are overlying at the instant of time $t = t' = 0$, from a point at the coinciding frame origins, in the relative motion direction. According to the light speed principle, after period of time t with respect to K, corresponding to t' with respect to K', has elapsed, the light ray tip will have travelled a distance $x = ct$

with respect to K, $x' = ct'$ with respect to K', where c
is the speed of light in empty space.

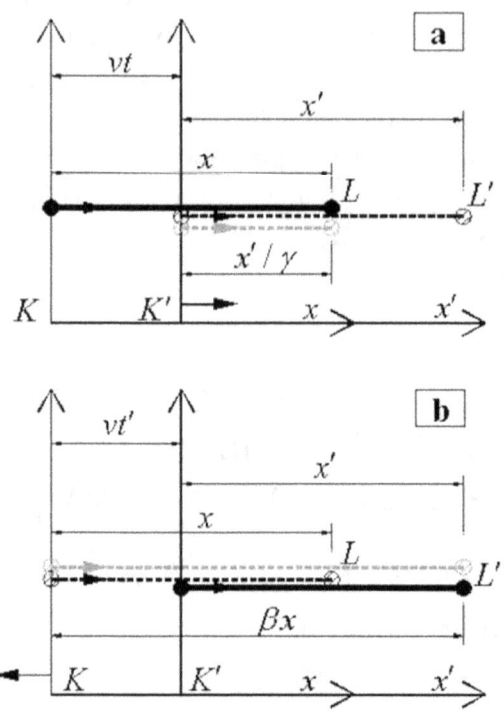

Fig. 4.1 Light ray tip point path from the perspective of K (a), and K' (b).

Since, according to the special relativity's second postulate, the speed of light is the same with respect to both frames, the light ray trajectory drawn independently in K and K' would appear as shown in Fig. 4.1 in solid lines with equal lengths. However, the light ray tip point L' is actually perceived as point L

Incompatibility of the Light
Speed Postulate with the Coordinate
Transformation Symmetry Assumption   95

(since L and L' represent the same event in each frame) with respect to K. Hence, the distance x' must be contracted with respect to K in order for point L' to coincide with point L. Suppose the distance x' is contracted by a factor of $1/\gamma$ (<1), as shown in Fig. 4.1a with the gray dashed line, the following expression is inferred from Fig. 4.1a, relative to K.

$$\gamma = \frac{x}{x - vt} = \frac{ct}{ct - vt} = \frac{1}{1 - \dfrac{v}{c}}. \qquad (4\text{-}1)$$

where vt is the distance travelled by K' with respect to K during the travel time t.

On the other hand, the light ray tip point L is actually perceived as point L', with respect to K' (since L and L' represent the same event in each frame). Hence, the distance x must then be expanded with respect to K' in order for point L to coincide with point L'. Suppose the distance x is expanded by the factor of $\beta > 1$, as shown in Fig. 4.1b with the gray dashed line. Hence, the following expression is inferred from Fig. 4.1b, relative to K'.

$$\beta = \frac{x' + vt'}{x'};$$

$$\frac{1}{\beta} = \frac{x'}{x' + vt'} = \frac{ct'}{ct' + vt'} = \frac{1}{1 + \dfrac{v}{c}}. \tag{4-2}$$

Eqs. (4-1) and (4-2) lead to

$$\frac{\gamma}{\beta} = \frac{1}{1 - \dfrac{v^2}{c^2}}. \tag{4-3}$$

If we now impose that the length in K must—by the "law" of symmetry—be also contracted with respect to K' by the factor $1/\gamma$ (i.e., $\beta = 1/\gamma$, contradicting the intrinsic length expansion relative to K'), then Eq. (4-3) reduces to

$$\gamma = \frac{1}{\sqrt{1 - \dfrac{v^2}{c^2}}}, \tag{4-4}$$

which is the Lorentz contraction factor, in accordance with to the special relativity predictions.

Consequently, comparing Eqs. (4-1) and (4-4), the symmetry requirement results in

Incompatibility of the Light
Speed Postulate with the Coordinate
Transformation Symmetry Assumption   97

$$\frac{1}{1-\dfrac{v}{c}} = \frac{1}{\sqrt{1-\dfrac{v^2}{c^2}}};$$

$$\left(1-\frac{v}{c}\right)^2 = 1-\frac{v^2}{c^2};$$

$$1-\frac{v}{c} = 1+\frac{v}{c}; \tag{4-5}$$

$$v = 0.$$

Or, the symmetry criteria $\beta = 1/\gamma$ leads to—from Eqs. (1) and (2)

$$1-\frac{v}{c} = 1+\frac{v}{c}, \text{ or } v = 0,$$

implying the reference frame must be at rest with respect to each other in order to satisfy the light speed principle and the transformation symmetry. It follows that the special relativity is deemed to be refuted.

COORDINATE TRANSFORMATION AND VERIFICATION OF FINDINGS

Using Fig. 4.1a, the following transformation is deduced from the perspective of an observer in K.

$$x = \frac{x'}{\gamma} + vt;$$
$$x' = \gamma(x - vt).$$
(4-6)

Similarly, Fig. 4.1b leads to the following transformation from the perspective of an observer in K'.

$$x' = \beta x - vt';$$
$$x = \frac{1}{\beta}(x' + vt').$$
(4-7)

Eqs. (4-6) and (4-7) lead to

$$\gamma x = x' + \gamma vt$$ (4-8)
$$\beta x = x' + vt'$$ (4-9)

Dividing Eq. (4-8) by Eq. (4-9) we obtain

Incompatibility of the Light
Speed Postulate with the Coordinate
Transformation Symmetry Assumption   99

$$\frac{\gamma}{\beta} = \frac{x'}{x' + vt'} + \frac{\gamma vt}{\beta x};$$

$$\frac{\gamma}{\beta} = \frac{1}{1 + \dfrac{v}{c}} + \frac{\gamma v}{\beta c};$$

$$\frac{\gamma}{\beta}\left(1 - \frac{v}{c}\right) = \frac{1}{1 + \dfrac{v}{c}};$$

$$\frac{\gamma}{\beta} = \frac{1}{1 - \dfrac{v^2}{c^2}},$$

$(4\text{-}10)$

verifying Eq. (4-3).

It should be noted that the spatial transformation Eqs. (4-6) and (4-7) deduced from the speed of light invariance are in conformance with the Galilean transformation for the limit $v \ll c$.

Now, if Eqs. (4-6) and (4-7) were to be symmetrical, in accordance with the special relativity assumption of transformation symmetry, then

$$\gamma = \frac{1}{\beta},$$

$(4\text{-}11)$

leading to (from Eq. (4-10))

$$\gamma = \frac{1}{\sqrt{1 - \frac{v^2}{c^2}}}, \qquad (4\text{-}12)$$

contradicting the intrinsic length expansion relative to K' (β becomes < 1), and Eqs. (4-6) and (4-7) become the spatial Lorentz transformation and its inverse. However, $\beta < 1$ would result in shifting point L to the opposite direction away from L' (Fig. 1b) making the same event occur in two different locations at the same time with respect to K', thus leading to an impossible occurrence. In fact, this would necessarily lead to (from Eqs. (4-1), (4-2), and (4-11))

$$v = 0.$$

It follows that the special relativity prediction is deemed to be refuted.

THE SPECIAL RELATIVITY BLUNDER

Using the isotropic property of space, and the Special Relativity first postulate stating that the laws of physics are the same in all inertial reference frames, the coordinate transformation with respect to the unprimed frame K, given by Eq. (4-6)—obtained from the constancy of the speed of light postulate—would

Incompatibility of the Light
Speed Postulate with the Coordinate
Transformation Symmetry Assumption   101

represent the inverse transformation (i.e., with respect to the primed frame K'), had we swapped in the equation the unprimed and the primed coordinates, and reverse the sign of the relative velocity (as K is traveling in the opposite direction with respect to K'). This will lead to the following transformation equation and its inverse.

$$x' = \gamma(x - vt) \qquad (4\text{-}13)$$

$$x = \gamma(x' + vt') \qquad (4\text{-}14)$$

Obviously, Eq. (4-14) is inconsistent with the speed of light principle, as it is not in line with Eq. (4-7) required by this principle.

Now, dividing both sides of Eqs. (4-13) and (4-14) by c, the speed of light, the following time transformation equations are obtained.

$$t' = \gamma t\left(1 - \frac{v}{c}\right) \qquad (4\text{-}15)$$

$$t = \gamma t'\left(1 + \frac{v}{c}\right) \qquad (4\text{-}16)$$

Substituting Eq. (4-15) into Eq. (4-16) leads after simple simplification to

$$\gamma = \frac{1}{\sqrt{1 - \dfrac{v^2}{c^2}}}. \qquad (4\text{-}17)$$

Replacing Eq. (4-16) in Eq. (4-13), and Eq. (4-15) in Eq. (4-14), returning, respectively

$$x = \gamma\left[x'\left(\frac{1}{\gamma^2} + \frac{v^2 t'}{cx'} \right) + vt' \right],$$

and

$$x' = \gamma\left[x\left(\frac{1}{\gamma^2} + \frac{v^2 t}{cx} \right) - vt \right],$$

requiring $x' = ct'$ and $x = ct$, to yield the transformation Eqs. (4-14) and (4-13), respectively. When this requirement (i.e., $x = ct$ and $x' = ct'$) is applied to Eqs. (4-15) and (4-16), the following equations are returned.

$$t' = \gamma\left(t - \frac{vx}{c^2} \right) \qquad (4\text{-}18)$$

$$t = \gamma\left(t' + \frac{vx'}{c^2} \right) \qquad (4\text{-}19)$$

Incompatibility of the Light
Speed Postulate with the Coordinate
Transformation Symmetry Assumption   103

It follows that, Eqs. (4-13), (4-14), (4-18), and (4-19), which are nothing but the Lorentz transformation equations, are restricted to $x = ct$ and $x' = ct'$, which leads to various contradictions.

In fact, when $t' = 0$, Lorentz transformation (4-18) leads to $t = vx / c^2$. But, as shown above, $x = ct$ in Eq. (4-18), yielding the contradiction $t = vct / c^2$, or $v = c$.

Similarly, Lorentz transformation (4-19) can lead to a similar contradiction for $t = 0$ (i.e., $v = -c$).

Furthermore, substituting Eq. (4-18) into Eq. (4-19) returns

$$t = \gamma \left(\gamma \left(t - \frac{vx}{c^2} \right) + \frac{vx'}{c^2} \right), \qquad (4\text{-}20)$$

which can be simplified to

$$t \left(\gamma^2 - 1 \right) = \frac{vx}{c^2} \left(\gamma^2 - \frac{\gamma x'}{x} \right). \qquad (4\text{-}21)$$

Since, as shown earlier, Eqs. (4-18) and (4-19) require $x = ct$; $x' = ct'$, then Eq. (4-21) can be written as

$$t \left(\gamma^2 - 1 \right) = \frac{vx}{c^2} \left(\gamma^2 - \frac{\gamma t'}{t} \right). \qquad (4\text{-}22)$$

Now, for time $t' = 0$, the transformed t-coordinate with respect to K would be $t = vx \,/\, c^2$, according to Eq. (4-18). Consequently, for $t \neq 0$, Eq. (4-22) would reduce to

$$t\left(\gamma^2 - 1\right) = t\gamma^2,$$

yielding the contradiction,

$$\gamma^2 - 1 = \gamma^2 , \quad \text{or} \quad 0 = 1 .$$

It follows that the conversion of the time coordinate $t' = 0$ to $t = vx \,/\, c^2$, for $x \neq 0$, by Lorentz transformation Eq. (4-18), is proved to be invalid, since it leads to a contradiction when used in Eq. (4-22), resulting from the Lorentz transformation equations for $t \neq 0$ (i.e., beyond the initial overlaid-frames instant satisfying $t = 0$ for $t' = 0$).

A similar contradiction is obtained by substituting Eq. (4-19) into Eq. (4-18), and applying Eq. (4-19) for the conversion $t = 0$; $t' = -vx' \,/\, c^2$.

In addition, substituting Eq. (4-13) into Eq. (4-14), yields

$$x = \gamma\left(\gamma\left(x - vt\right) + vt'\right);$$
$$x\left(\gamma^2 - 1\right) = \gamma v\left(\gamma t - t'\right);$$

Incompatibility of the Light
Speed Postulate with the Coordinate
Transformation Symmetry Assumption   105

$$x\left(\gamma^2 - 1\right) = \gamma vt\left(\gamma - \frac{t'}{t}\right). \qquad (4\text{-}23)$$

Since Eqs. (4-13) and (4-14)—along with Eqs. (4-18) and (4-19)—require $x = ct$; $x' = ct'$, Eq. (4-23) can be written as

$$x\left(\gamma^2 - 1\right) = \gamma vt\left(\gamma - \frac{x'}{x}\right). \qquad (4\text{-}24)$$

Now, for $x' = 0$, the transformed x-coordinate with respect to K would be $x = vt$, according to Eq. (4-13). Consequently, for $x \neq 0$, Eq. (4-24) would reduce to

$$x\left(\gamma^2 - 1\right) = x\gamma^2,$$

$$\gamma^2 - 1 = \gamma^2, \quad \text{or} \quad 0 = 1.$$

It follows that the conversion of the space coordinate $x' = 0$ of K' origin to $x = vt$, at time $t > 0$, with respect to K by Lorentz transformation equation, is invalid, since it leads to a contradiction when used in Eq. (4-24), resulting from Lorentz transformation equations, for $x \neq 0$ (i.e., beyond the initial overlaid-frames position satisfying $x = 0$ for $x' = 0$).

A similar contradiction would follow upon substituting Eq. (4-14) into Eq. (4-13), and applying Eq. (4-14) for the conversion $x = 0$; $x' = -vt'$.

CONCLUSION

Considering two internal reference frames —unprimed and primed—in relative translational motion, the direct coordinate conversion factor and its inverse were easily deduced from the constancy of the speed of light principle, using simple diagrams for a light ray travel path from the perspective of each of the two frames. The direct length conversion factor was found to be in agreement with the corresponding special relativity prediction. However, the deduced inverse conversion factor was not symmetrical with respect to the direct length conversion that required that the space in the primed frame be contracted with respect to that of the unprimed frame, while the inverse length conversion factor showed the inverse relation (i.e., the length in the unprimed frame was expanded with respect to the primed frame). It followed that, to achieve symmetry (i.e., length be mutually contracted with respect to both frames and by the same factor), the constancy of the speed of light principle required that the two frames be at rest with respect to each other, thus invalidating the special relativity predictions. Moreover, further analysis of the Lorentz transformation, following from the coordinate

Incompatibility of the Light
Speed Postulate with the Coordinate
Transformation Symmetry Assumption   107

transformation symmetry assumption, showed fatal mathematical contradictions leading to its refutation.

Chapter 5

INCONSISTENCY OF THE SPECIAL RELATIVITY WITH THE PRINCIPLE OF RELATIVITY

The fact that the length of an object, relative to its rest frame, is independent of the object's spatial orientation always holds and cannot be violated or demonstrated to be otherwise by any experimental or analytical means. Therefore, this fact becomes a principle that, by definition, can be brought to the level of a physics law. However, according to the Special Relativity (SR) this law is not applicable in an inertial reference frame in relative motion with respect to the object, thus violating the principle of relativity, the SR first postulate that states: "The laws of physics are the same in all inertial frames of reference". This violation leads to many contradictions between the SR and its first postulate. Indeed, simple thought experiments in different areas of physics are examined showing how the SR predictions result in outcomes inconsistent with the principle of relativity. Relatively moving apparatuses physical measurements such as pressure vessel discharge time, optical lens focal length, and electrical resistance are shown—according to SR—to vary with the respective apparatus orientation in the space relative to the stationary reference frame. This is merely a violation of the relativity principle, since these measurements, in line with the applicable physics laws they're subjected to, are independent of the apparatuses orientation with respect to the reference frames they travel with (i.e., with respect to their rest frames).

BACKGROUND

The laws of physics govern the description of all physical aspects of an object, like its mass, dimensions, and state, as well as its thermal, electrical, and optical properties. If these governing laws of physics were somehow altered, the object will cease to have the same physical aspects. Particularly, the fact that the length of an object is a physical aspect that is independent of the object's orientation in the space relative to the object's rest frame can be considered, by definition, a physics law—it always holds, and cannot be proved to be otherwise by any means. It follows that, when the relativity principle talks of invariant laws of physics in all inertial reference frames, it also means the physical aspects of an object under certain physical conditions in an arbitrary inertial reference frame must be the same from the perspective of any other inertial frame of reference. However, with the predictions of SR, where the physical length of an

object is altered by its relative motion depending on
the object's spatial orientation with respect to the
"stationary" reference frame, this principle of relativity
cannot hold, creating a predicament as the SR is
developed relying on this principle. In this Chapter, the
inconsistency of the SR with the principle of relativity
is demonstrated through thought experiments, dealing
with different areas of physics. These experiments
show how altered length dimensions, as predicted by
the SR, generate results in contradiction with the
principle of relativity.

THOUGHT EXPERIMENTS

Pressure Vessel Discharge Time

Let $K(x, y, z, t)$ be a coordinate system related to an
observer's reference frame K in the outer space.
Consider a pressurized cylindrical gas vessel located in
another reference frame K' (e.g., a space craft) (Fig.
5.1), in uniform translational motion with respect to K,
attached to it a coordinate system $K'(x', y', z', t')$, with
the coordinate axes being parallel to the corresponding
ones in K. Let the relative velocity v be in the
common x-x' direction. Let V denote the vessel's
volume, P_o and T its initial absolute pressure and
temperature (equal to surrounding's), respectively, all
with respect to K'. Let the surrounding space pressure

be P_s. The vessel is fitted with two identical circular nozzles, of cross sectional area A, connected to the vessel through isolating valves. The nozzles are mounted in such a way they discharge in opposite directions, so that the resultant of their thrust forces is zero, with their cross section planes being parallel to the y'-axis. The valves are then opened so as to let the pressurized gas discharge isothermally to the surrounding, bringing the vessel gauge pressure to zero (i.e., bringing the absolute pressure to surrounding pressure) at constant temperature. The objective is to calculate the gas discharge time duration using the applicable laws of physics. Accordingly, this duration, independent of the vessel orientation, can be measured in K' using the formula

$$\Delta t' = \frac{VM}{RTCA}\sqrt{\Gamma(k)\frac{P_o - P_s}{\rho_o}}. \qquad (5\text{-}1)$$

Where,
$\Delta t'$ = vessel's discharge time
V = vessel's volume
M = molar mass of the gas
R = universal gas constant
T = absolute temperature
C = flow coefficient (dimensionless)
A = nozzles cross sectional area

$\Gamma(k) =$ dimensionless parameter, a function of the gas specific heat ratio $k = c_p / c_v$

$P_o =$ initial gas pressure in the vessel

$P_S =$ surrounding pressure

$\rho_o =$ initial gas density

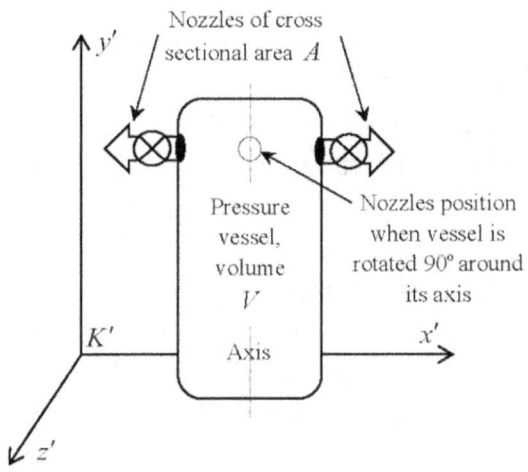

Fig. 5.1 Experimental arrangement

Special Relativity Prediction

From the perspective of K, according to SR[1], the diameter of the vessel along the x-direction is contracted by the relativistic factor of γ[a], thus its cross

---

[a] $\gamma = 1 / \sqrt{1 - v^2 / c^2}$, where c is the speed of light in empty space.

sectional area as well as its volume are contracted by the same factor, given its dimensions in the y - and x - directions remain invariant.

Whereas, the gas mass—as well as its molar mass—is increased (scaled) by the factor γ, hence its density is scaled by γ^2. According to the gas laws for an isothermal process, the vessel's initial gas pressure would then be augmented (scaled) by the same volume contraction factor γ.

As for the nozzles cross sectional area, whether it's affected by the relative motion depends on the vessel's orientation and the respective nozzles position with respect to the relative motion direction.

Case 1 — the nozzle cross section planes are parallel to the yz plane, i.e. perpendicular to the motion direction: Here, according to SR predictions, the nozzles diameter remains invariant, and, in accordance with the relativity principle, the discharge time duration with respect to K would be measured as

$$\Delta t = \frac{(V/\gamma)\gamma M}{RTCA}\sqrt{\Gamma(k)\frac{\gamma P_o - P_s}{\gamma^2 \rho_o}};$$

$$(5\text{-}2)$$

$$\Delta t = \frac{VM}{\gamma RTCA}\sqrt{\Gamma(k)\frac{\gamma P_o - P_s}{\rho_o}},$$

Case 2 — the nozzle cross section planes are parallel to the xy plane, i.e. parallel to the motion direction: Here, in accordance with SR, the nozzles diameter along the x-direction is contracted by γ, hence their cross sectional area is also contracted by the same factor, whereas all the other parameters remain unchanged, and, in accordance with the relativity principle, the discharge time duration with respect to K would be measured as

$$\Delta t = \frac{VM}{RTCA}\sqrt{\Gamma(k)\frac{\gamma P_o - P_s}{\rho_o}}, \qquad (5\text{-}3)$$

returning a longer duration than Eq. (5-2) in case 1, thus contradicting the physics laws according to which the discharge duration as given in Eq. (5-1) with respect to K', should be independent of the orientation of the vessel in the space in all reference frames. In fact, this unrealistic SR outcome implies that as the cylinder is rotated around its axis in K', the gas discharge time will undergo changes with respect to K, since the nozzles cross sectional area varies according to their orientation with respect to the relative motion direction!

Optical Lens Properties

In this thought experiment, we keep our previous frames of reference, and we replace the vessel in K' with an optical convergent lens made of glass of refractive index n. The lens longitudinal axis is parallel to the y'-axis. The spherical boundary surfaces form a variable lens thickness along the longitudinal axis. Let d be the lens thickness at its optical center, and R the radius of each of the spherical boundaries. The objective is to measure the focal length of the lens, using the applicable laws of physics. Accordingly, this focal length, independent of the lens orientation, can be measured with respect to K' using the known formula

$$\frac{1}{f'} = \frac{(n-1)^2 d}{nR^2}.$$
(5-4)

Special Relativity Prediction

From the perspective of K, according to SR, whether the lens thickness would contract depends on the lens optic axis direction with respect to the relative motion direction.

Case 1— the optic axis is parallel to the x-direction: In this case, according to SR, the thickness of the lens is contracted by the relativistic factor γ at any point along the lens longitudinal axis. Consequently, the radii of the bounding surfaces will be extended (scaled) by the factor γ (i.e., inversely proportional to the thickness), since the lens curvatures are decreased.

In addition, since the lens volume is proportional to the thickness, it will be reduced (divided) by γ, causing the lens density to be multiplied by γ^2, since according to SR the mass is also scaled by γ, hence its index of refraction would be scaled by γ, had we assumed it was proportional to the square root of the material's density for the used glass. Therefore, in accordance with the relativity principle, the lens focal length with respect to K would be measured as

$$\frac{1}{f} = \frac{(\gamma n - 1)^2 d}{\gamma^4 n R^2},$$
(5-5)

returning a longer focal length than Eq. (5-4) with respect to K', which is in contradiction with the relativity principle, since the same governing physics laws in both frames (K and K') should result in the same focal length, a physical aspect of the lens.

This focal length increase has a fatal consequence on the SR. Indeed, a screen-focused lens image relative to K' would become out of focus with respect to K, since, according to SR, the relative motion causes the focal length to vary in opposite trend to the space dimension in the motion direction (i.e., while the space dimension is contracted with respect to K, the focal length is extended).

Case 2— the optic axis is parallel to the *z*-direction: In this case, according to SR, the thickness of the lens remains unchanged. However, the lens width will be contracted by γ, resulting in volume decrease, hence an increase in the lens density and, as in case 1, in its refraction index, whereas all the other parameters remain unchanged, and, in accordance with the relativity principle, the formula for measuring the focal length f with respect to K becomes

$$\frac{1}{f} = \frac{(\gamma n - 1)^2 d}{\gamma n R^2},\qquad (5\text{-}6)$$

returning a shorter focal length (higher lens power) than Eq. (5-5) in case 1—and Eq. (5-4) with respect to K'—thus contradicting the physics laws according to which the lens focal length—a physical property governed by the laws of physics—as given in Eq. (5-4) with respect to K' must be independent of the orientation of the lens in the space in all reference frames. In fact, this unrealistic SR outcome implies that, with respect to K, as an object-lens set is rotated in K' around the lens longitudinal axis, the formed image will undergo changes in size and distance from the lens, since the focal length varies according to the optic axis orientation with respect to the relative motion direction!

Electrical Resistance

Here, a cylindrical object made of a certain material of electrical resistivity ρ is considered in the reference frame K'. The object has a length l, and a cross sectional area A. With respect to K', the electrical resistance of this object can be measured from the formula

$$R' = \rho\left(\frac{l}{A}\right). \qquad (5\text{-}7)$$

Special Relativity Prediction

From the perspective of K, according to SR, the dimensions of the object are altered depending on its orientation with respect to the direction of the relative motion.

Case 1 — the cylinder axis is parallel to the x-direction: In this case, according to SR, the length of the object is contracted by the relativistic factor γ, whereas its cross sectional area remains the same. Hence, the resistance with respect to K becomes

$$R = \frac{\rho}{\gamma}\left(\frac{l}{A}\right), \qquad (5\text{-}8)$$

returning a lower resistance than Eq. (5-7), which is in contradiction with the relativity principle, since the same governing physics laws in both frames should result in the same electrical resistance, a physical aspect of the object.

Case 2 — the cylinder axis is perpendicular to the x-direction: Here, according to SR, the length of the object is unaltered by the relative motion, whereas its cross sectional area is contracted by the relativistic factor γ. Hence, the resistance with respect to K would be changed to

$$R = \rho\gamma\left(\frac{l}{A}\right), \qquad (5\text{-}9)$$

returning a higher resistance than Eq. (5-8) in case 1 (scaled by γ^2)—and scaled by γ relative to the resistance in K'—thus contradicting the physics laws according to which the electrical resistance—a physical property governed by the applicable physics laws—as given in Eq. (5-7) with respect to K' must be independent of the orientation of the object in the space in all reference frames. In fact, this unrealistic SR outcome implies that as an electric circuit in the $x'z'$ plane is rotated in K' around an axis perpendicular to the relative motion direction, the circuit current will undergo changes in magnitude with respect to K, since the circuit resistance varies according to its elements orientation with respect to the relative motion direction!

Thermal Conductivity

A similar contradiction could be concluded from the SR predictions with respect to the conduction heat transfer through an object, in terms of exhibiting different thermal conductivities in different inertial reference frames in relative motion, and in different orientations in the space with respect to the "stationary" reference frame.

CONCLUSION

It follows that the SR predictions result in contradictory outcomes when applied in accordance with the relativity principle. This is simply attributed to the fact that the length of an object, being a physical dimension, is a physical property governed by the pertinent laws of physics according to which the object's length is independent of its spatial orientation in its rest frame; hence, according to the relativity principle, an object must have the same physical dimensions with respect to all inertial frames of reference, regardless of its spatial orientation, which is in contradiction with the SR predicting length variations of an object traveling with an inertial reference frame in relative motion, depending on its orientation with respect to the stationary reference frame.

Chapter 6

RECONFIRMATION OF THE CRITICAL ERROR IN THE FORMULATION OF THE SPECIAL RELATIVITY

The perception of events in two inertial reference frames in relative motion is analyzed from the perspective of the special relativity (SR) postulates. Straightforward contradictions disproving the SR predictions have been identified. The hoax used in the SR formulation to get around the identified contradictions and hide their cause is revealed.

BACKGROUND

The SR time dilation prediction is based on the transformation of the time interval between two events occurring at the origin (or co-local events) in one reference frame to another frame in relative translational motion with respect to the first. It has been shown in chapters 2 and 3 that such transformation would be invalid, as it involves coordinates having zero value. In this chapter, further analysis of event perceptions relative to both frames will reconfirm the invalidity of such transformation, hence the invalidity of the SR prediction of the time dilation. Similar analysis for simultaneous events proves the invalidity of the SR length contraction prediction.

TEMPORAL EVENTS ANALYSIS

Consider two inertial frames of reference, $K(x,y,z,t)$ and $K'(x',y',z',t')$, in translational relative motion with parallel corresponding axes, and let their origins be aligned along the overlapped x- and x'-axes. Let v be the relative motion velocity in the x-x' direction. K and K' are assumed to be overlapping at the time $t = t' = 0$.

Arbitrary non-origin events

Let's suppose that at the frames overlapping instant, an event $E_1(x',0,0,0)$ $[E_1(x,0,0,0)]$ takes place at distance x' with respect to K' origin (x with respect to K origin) on the x-x' axis. According to the SR light speed postulate, this event is perceived by an observer at K' origin at the time

$$t' = \frac{x'}{c}, \qquad (6\text{-}1)$$

and by an observer at K origin at the time

$$t = \frac{x}{c}. \qquad (6\text{-}2)$$

With respect to the K' observer, the origin of K' at this perception time is at a distance of vt' from that of K, and the same event will be perceived (with respect to the K' observer) by an observer at K origin at the time

$$t = t' + \frac{vt'}{c}. \qquad (6\text{-}3)$$

To account for any time scaling due to the relative motion between the inertial frames K and K', let's write Eq. (6-3) in the following form.

$$t = \gamma\left(t' + \frac{vt'}{c}\right), \qquad (6\text{-}4)$$

where γ is a real positive factor depending on v.
Replacing Eq. (6-1) into Eq. (6-4) yields

$$t = \gamma\left(t' + \frac{vx'}{c^2}\right). \qquad (6\text{-}5)$$

Multiplying both sides of Eq. (6-4) by c, and using Eqs. (6-1) and (6-2) leads to

$$x = \gamma(x' + vt'). \qquad (6\text{-}6)$$

As demonstrated in chapters 2 and 3, by using the SR first postulate the inverse of the transformation Eqs. (6-5) and (6-6) can be obtained by swapping the primed and unprimed coordinates, and replacing v with $-v$. Solving Eqs. (6-5), (6-6) and the resulting inverse transformation equations for γ results in

$$\gamma = \frac{1}{\sqrt{1 - \frac{v^2}{c^2}}}. \qquad (6\text{-}7)$$

Eqs. (6-5) and (6-6) are therefore the [inverse] Lorentz transformation equations for the coordinates in the relative motion direction.

Co-local Events at K' origin

Now, suppose an event $E_2(0,0,0,t')$ $[E_2(vt,0,0,t)]$ occurs at K' origin,

$$x' = 0, \tag{6-8}$$

at the time t' with respect to K' (t with respect to K).

Again, with respect to the K' observer, the origin of K' at the event perception time is at a distance of vt' from that of K, and the same event will be perceived (with respect to the K' observer) by an observer at K origin at the time

$$t = t' + \frac{vt'}{c}, \tag{6-9}$$

or, to account for any time scaling, at the time

$$t = \gamma \left(t' + \frac{vt'}{c} \right). \tag{6-10}$$

However, in this case Eq. (6-1) doesn't hold, and therefore Eq. (6-5) doesn't follow. Yet, in SR it is customary for such events (occurring at K' origin) to replace Eq. (6-8) ($x' = 0$) in Lorentz transformation Eq. (6-5), inapplicable in this case, since it is derived for events having $x' = ct'$ invalid for co-local events having $x' = 0$ and $t' > 0$.

Therefore, for an event occurring at K' origin ($x' = 0$) at time t', the SR-predicted time t with respect to K is concluded from the invalid (for this case) Eq. (6-5) as

$$t = \gamma t'. \tag{6-11}$$

Whereas, Eq. (6-10) predicts this time to be

$$t = \gamma t' \left(1 + \frac{v}{c}\right). \tag{6-12}$$

Comparing Eqs. (6-11) and (6-12) results in the contradiction

$$v = 0. \tag{6-13}$$

It follows that the SR conversion $x' = 0;\ t = \gamma t'$, predicting time dilation, is invalid.

The same analysis of the above two events can be performed from the perspective of an observer at K origin, with a similar contradiction being obtained.

Simultaneous events

Similarly, Lorentz transformation Eq. (6-6) is not applicable for events having $t' = 0$ and $x' \neq 0$, as it is derived under Eqs. (6-1) and (6-2), requiring $x' = 0$ for $t' = 0$. However, in SR interpretation of Lorentz transformation Eq. (6-6), length contraction is predicted by setting $t' = 0$ (for simultaneous events duration) to get the relation $x = \gamma x'$, ignoring the restriction imposed by the basic speed of light constancy Eqs. (6-1) and (6-2). Hence follows the invalidity of the SR length contraction prediction.

THE SPECIAL RELATIVITY HOAX

It is ascertained in the previous sections that the Lorentz transformation time equations

$$t' = \gamma\left(t - \frac{vx}{c^2}\right),$$

$$t = \gamma\left(t' + \frac{vx'}{c^2}\right),$$

are principally derived on the basis of events having $x = ct$; $x' = ct'$, invalid for co-local events having $x = 0$ and $t > 0$ $(x' = 0$ and $t' > 0)$. These restrictions are obviously fatal for the SR formulation requiring such co-local events—separated by a time interval—for the interpretation of the Lorentz transformation. In order to overcome this obstacle, the equations

$$x = ct, \tag{6-14}$$

and

$$x' = ct', \qquad (6\text{-}15)$$

expressing the basic speed of light constancy principle, were manipulated and combined into the equation

$$x^2 - c^2 t^2 = x'^2 - c^2 t'^2, \qquad (6\text{-}16)$$

set as the principle equation representing the SR speed of light postulate.[5,19] Setting $x = 0$ with $t > 0$ (or $x' = 0$ with $t' > 0$); or $t = 0$ with $x \neq 0$ (or $t' = 0$ with $x' \neq 0$), is made now possible with the constructed Eq. (6-16), while the conditions $x = 0$; $t = 0$ ($x' = 0$; $t' = 0$), imposed by the original light speed constancy Eqs. (6-14) and (6-15), are ignored!

It should be noted that Eq. (6-16) can also be obtained from the light sphere equations, namely

$$x^2 + y^2 + z^2 = c^2 t^2, \qquad (6\text{-}17)$$

$$x'^2 + y'^2 + z'^2 = c^2 t'^2, \qquad (6\text{-}18)$$

representing the light speed constancy principle in the three-dimensional space, by subtracting the two equations from each other, and using the invariance of the y and z coordinates (i.e., $y = y'$, $z = z'$). However, Eqs. (6-17) and (6-18) also require that at the instant of time $t = t' = 0$—the moment when the spherical light wave front is emitted from the coinciding frame origins—the spatial coordinates must be zero as well, i.e., $x = x' = 0$, $y = y' = 0$, and $z = z' = 0$; these initial conditions are not attributed to the resulting Eq. (6-16) in the SR formulation.

Eq. (6-16) forms the basis of the Lorentz transformation derivation in the SR formulation.[5] The Lorentz transformation equations are indeed derivable, yet more tediously, from Eq. (6-16) being mathematically equivalent to the deriving Eqs. (6-14) and (6-15)—except with no consideration given to the coordinate values obtained from these equations at the space and time origins (i.e., ignoring the initial conditions required by Eqs. (6-14) and (6-15)). Such a critical violation undermines the validity of the SR predictions, in agreement with the findings of earlier studies.[20,21] In fact, these studies demonstrate that the Lorentz transformation equations result in mathematical contradictions when applied for co-local or simultaneous events.

CONCLUSION

Once again, the Lorentz transformation equations are shown to be merely applicable for events satisfying the basic light speed constancy equations $x = ct$ and $x' = ct'$. The erroneous application of the Lorentz transformation on co-local events $(x' = 0;\ t' > 0,$ in K', or $x = 0;\ t > 0,$ in $K)$, or simultaneous events $(t' = 0;\ x' \neq 0,$ in $K',$ or $t = 0;\ x \neq 0,$ in $K)$, is shown to result in invalid predictions of time dilation, or length contraction, respectively.

APPENDIX A

Derivation of the Lorentz Transformation

The Lorentz transformation was derived by Einstein[1,5] on the basis of the relativity principle and the constancy of the speed of light postulate. The sought transformation, converting between the space and time coordinates of two inertial reference frames, say $K(x,y,z,t)$ and $K'(x',y',z',t')$, in relative motion at speed v along the x-x' axis, is assumed to take the following general linear form:

$$x' = ax + bt \qquad (A.1)$$
$$y' = y, \; z' = z \qquad (A.2)$$
$$t' = kx + mt \qquad (A.3)$$

where a, b, k, and m are unknown real terms to be determined.

On the other hand, the constancy of the speed of light postulate is expressed by the assumption that a spherical light wave front, emitted from the coinciding inertial reference frame origins, would be observed as a light sphere centered at the frame origin, with its radius being expanded at the speed of light c, with respect to either frame:

$$x^2 + y^2 + z^2 = c^2 t^2,$$

$$x'^2 + y'^2 + z'^2 = c^2 t'^2.$$

Using Eq.(A.2), yielding $y^2 + z^2 = y'^2 + z'^2$, the above light sphere equations lead to

$$x'^2 - c^2 t'^2 = x^2 - c^2 t^2, \qquad (A.4)$$

the basic constancy of the speed of light equation.

Plugging the particular conversion $x' = 0$; $x = vt$ in Eq. (A.1) yields the relation

$$0 = avt + bt,$$

or (for $t \neq 0$),

$$b = -av,$$

leading to

$$x' = a\left(x - vt\right). \qquad (A.5)$$

Now, owing to the fact that the reference frame K is traveling at a speed of $-v$ with respect to K', and to the essential symmetrical property—inferred from the principle of relativity—of the transformation with respect to the reference frames, the inverse of the transformation given by Eq. (A.5) can be written as

$$x = a\left(x' + vt'\right), \qquad (A.6)$$

which must also hold—by symmetry—for t' ≠ 0 .

Solving Eqs. (A.5) and (A.6) for t', we get

$$t' = a\left[t - \frac{x}{v}\left(1 - \frac{1}{a^2}\right)\right]. \qquad (A.7)$$

Therefore, Eqs. (A.3) and (A.7) yield

$$m = a,$$

and

$$k = -\frac{a}{v}\left(1 - \frac{1}{a^2}\right).$$

It follows that we are left with determining the term ' a ' to obtain the final shape of the proposed transformation described by Eqs. (A.1) and (A.3), being reduced to Eqs. (A.5) and (A.7).

Substituting Eqs. (A.5) and (A.7) into the basic constancy of the speed of light equation Eq. (A.4), and equating the coefficients of x^2, t^2 , and xt on both sides of the resulting equation, the following expressions are obtained.

$$a^2 - \frac{c^2 a^2}{v^2}\left(1 - \frac{1}{a^2}\right)^2 = 1,$$

$$a^2 c^2 - a^2 v^2 = c^2,$$

and

$$\frac{2c^2 a^2}{v}\left(1 - \frac{1}{a^2}\right) - 2va^2 = 0.$$

Any of the above expressions will equally lead to the value of a, being

$$a = \frac{1}{\sqrt{1 - \dfrac{v^2}{c^2}}} = \gamma.$$

Replacing a, b, k, and m with their obtained values in Eqs. (A.1) and (A.3), yields the Lorentz transformation equations:

$$x' = \gamma\left(x - vt\right)$$

$$t' = \gamma\left(t - \frac{vx}{c^2}\right)$$

$$y' = y$$

$$z' = z$$

$$\gamma = \frac{1}{\sqrt{1 - \dfrac{v^2}{c^2}}}$$

REFERENCES

1   Einstein, A. Zur elektrodynamik bewegter Körper. *Annalen der Physik* **322**, 891–921 (1905).

2   Michelson, A. A. & Morley, E. H. On the Relative Motion of the Earth and the Luminiferous Ether. *Am. J. Sci.* **34**, 333-345 (1887).

3   FitzGerald, G. F. The Ether and the Earth's Atmosphere. *Science* **328** 390 (1889).

4   Lorentz, H. A. The Relative Motion of the Earth and the Aether. *Zittingsverlag Akad. V. Wet.* **1** 74–79 (1892).

5   Einstein, A. Einstein's comprehensive 1907 essay on relativity, part I. *English translations in Am. Jour. Phys. 45 (1977), Jahrbuch der Radioaktivitat und Elektronik* **4** (1907).

6   Beckmann, P. *Einstein Plus Two.* (Golem Press, 1987).

7   McCausland, I. The Persistent Problem of Special Relativity. *Physics Essays* **18** (2005).

8   Hatch, R. *Escape from Einstein* (Kneat Kompany 1992).

9   Selleri, F.   (Apeiron Montreal, 1998).

10   Dingle, H. *Science at the Crossroads* (Martin Brian and O'keeffe, 1972).

11   Monti, R. A. Theory of Relativity : a critical analysis. *Physics Essays* **9**, 238 (1996).

12   Kelly, A. *Challenging Modern Physics: Questioning Einstein's Relativity Theories* (Brown Walker Press, 2005).

13   Poincaré, H. On the Dynamics of the Electron
      Comptes Rendus **140**, 1504–1508 (1905).
14   Wang, L. J. in *Physics and Modern Topics in
      Mechanical and Electrical Engineering*   (ed
      Nikos Mastorakis)   45 (World Scientific and
      Engineering Society Press, 1999).
15   Kassir, R. M. Inconsistency of the Special
      Relativity with the Principle of Relativity.
      *International Journal of Physics and
      Astronomy* **1**, 30-34 (2013).
16   Hafele, J. C. & Keating, R. E. Around-the-
      World Atomic Clocks: Predicted Relativistic
      Gains. *Science* **177**, 166-168 (1972).
17   Rossi, B., Hilberry, N. & Hoag, J. B. The
      variation of the hard component of cosmic rays
      with height and the disintegration of mesotrons.
      Phys. Rev. **57**, 461-469 (1940).
18   Rossi, B. & Hall, D. B. Variation of the Bate of
      Decay of Mesotrons with Momentum. *Phys.
      Rev.* **59**, 223 (1941).
19   Einstein, A. *Relativity: The Special and the
      General Theory* (Routledge, 1916).
20   Kassir, R. M. On Lorentz Transformation and
      Special Relativity: Critical Mathematical
      Analyses and Findings. *Physics Essays* **27**, 16
      (2014).
21   Kassir, R. M. On Special Relativity: Root cause
      of the problems with Lorentz transformation.
      Physics Essays **27**, 198-203 (2014).

INDEX

About the Author

Radwan M. Kassir, C.Eng, is a professional consultant engineer. He was educated at Northern Arizona University and Arizona State University where he received his Master of Science in Engineering. Down his career, Radwan has taught college-level courses in physics, particularly kinematics, dynamics, and modern physics, as well as courses in fluid dynamics and heat transfer. He has several published papers on relativity in Physics Essays and the International Journal of Physics & Astronomy. He earned the credentials of the LEED AP BD+C from the U.S. Green Building Council, and the PMP from the Project Management Institute. Radwan is a member of the Chartered Institute of Building Services Engineers (CIBSE).